FOUNDATIONS OF
GEOGRAPHIC INFORMATION SCIENCE

FOUNDATIONS OF
GEOGRAPHIC INFORMATION SCIENCE

Edited by

Matt Duckham, Michael F. Goodchild and
Michael F. Worboys

CRC Press
Taylor & Francis Group
Boca Raton London New York

CRC Press is an imprint of the
Taylor & Francis Group, an **informa** business
A TAYLOR & FRANCIS BOOK

CRC Press
Taylor & Francis Group
6000 Broken Sound Parkway NW, Suite 300
Boca Raton, FL 33487-2742

First issued in paperback 2020

© 2003 by Taylor & Francis Group, LLC
CRC Press is an imprint of Taylor & Francis Group, an Informa business

No claim to original U.S. Government works

ISBN-13: 978-0-367-45459-3 (pbk)
ISBN-13: 978-0-415-30726-0 (hbk)

Visit the Taylor & Francis Web site at
http://www.taylorandfrancis.com

and the CRC Press Web site at
http://www.crcpress.com

British Library Cataloguing in Publication Data
A catalogue record for this book is available from the British Library

Library of Congress Cataloging-in-Publication Data
A catalog record for this book has been requested

CONTENTS

CONTRIBUTORS

Harith Alani, Electronics and Computer Science Department, University of Southampton, Southampton, SO17 1BJ, UK

Thomas Bittner, Institute for Formal Ontology and Medical Information Science, University of Leipzig, Leipzig D-04107, Germany
and
Department of Computer Science, McCormick School of Engineering, Northwestern University, Evanston, IL 60201, USA

Andrew U. Frank, Institute for Geoinformation, Technical University of Vienna, Vienna, A-1040, Austria

Antony Galton, School of Engineering and Computer Science, University of Exeter, Exeter, EX4 4PT, UK.

Michael F. Goodchild, Geography Department, University of California, Santa Barbara, CA 93106, USA

Christopher Habel, Department for Informatics, University of Hamburg, Hamburg D-22527, Germany

Stephen C. Hirtle, School of Information Sciences, University of Pittsburgh, Pittsburgh, PA 15260, USA

Christopher B. Jones, Department of Computer Science, Cardiff University, Cardiff, CF24 3XF, UK

Ray R Larson, School of Information Management and Systems, University of California, Berkeley, CA 94720, USA

David M. Mark, Department of Geography, University of Buffalo, Buffalo, NY 14261, USA

Daniel R. Montello, Geography Department, University of California, Santa Barbara, CA 93106, USA

Barry Smith, Institute for Formal Ontology and Medical Information Science, University of Leipzig, Leipzig D-04107, Germany
and
Center for Cognitive Science, Baldy Hall, University at Buffalo, Buffalo, NY 14260, USA

John G. Stell, School of Computing, University of Leeds, Leeds, LS2 9JT, UK

Sabine Timpf, Department of Geography, University of Zürich, Zürich, CH-8057, Switzerland

Douglas Tudhope, School of Computing, University of Glamorgan, Pontypridd, CF37 1DL, UK

Michael F. Worboys, National Center for Geographic Information and Analysis, University of Maine, Orono, ME 04469-5711, USA

Introduction

This book contains a collection of chapters by leading researchers on the topic of geographic information science.

> Information science can be defined as the systematic study according to scientific principles of the nature and properties of information. Geographic information science is the subset of information science that is about geographic information. (Goodchild, 1992)

The focus on 'the I in GI Science provided an impetus for a meeting of 17 invited international researchers in the field, that took place in Manchester, England in the summer of 2001. The meeting coincided with the award by Keele University of an honorary doctorate to Michael Goodchild. The brief that we provided participants asked that each should provide in advance of the meeting a substantial paper that would address in some way the issue of the fundamental nature of geographic information. Specific questions were given, to act as prompts to participants and to set the general agenda. These questions were:

- What is geographic information?

- What fundamental principles are associated with geographic information?

- How can geographic information be represented?

- How does geographic information represent the world?

- How can geographic information be quantified, and what metrics of geographic information measure its volume?

- How can geographic information be communicated?

- What are the fundamental human-computer interaction issues associated with geographic information?

The resulting papers were placed on a web site in advance of the Manchester meeting, and then presented and discussed by the participants. Subsequently, the view of participants was that the papers should be made into a collection, after substantial revision in the light of discussions at Manchester and authors readings of the of other papers. This book is the result of that process. The editors have attempted to present the papers in a useful sequence, beginning with the scope of the field (Chapter 1), discussing the foundations in information science (Chapters 2–4) and cartographic representation (Chapter 5), considering the role of granularity (Chapter 6–7), dealing with key geographic information science concepts of boundary, region, neighborhood, and landmark (Chapters 8–10), discussing issues pertinent to digital geo-libraries (Chapters 11–12), and concluding with a paper motivated by human-computer interaction issues (Chapter 13).

My role was to dream up the idea and set the structure for the meeting and subsequent book. I would like to take this opportunity of thanking my fellow editors, Michael Goodchild and Matt Duckham. Without Mike, we would have been unable to attract such an illustrious cast to Manchester. Everyone was keen to take part in an indirect way in the honor that Keele was to bestow upon him. Mike provided general guidance throughout. Matt put an enormous amount of effort into making the meeting go smoothly, and is almost solely responsible for the very fine editorial work that went into this book. We three would all like to thank Taylor and Francis, and in particular Tony Moore and Sarah Kramer, for supporting the meeting and providing valuable assistance in the preparation of this book.

As David Mark concludes in Chapter 1, geographic information science is a fuzzy concept, but one with 'considerable depth, and a richness of intellectual challenges that mark it as a legitimate multidisciplinary field and perhaps an emerging new discipline'. It is to be hoped that this book will contribute to that possible emergence.

Mike Worboys
Maine, August 2002.

REFERENCES

Goodchild, M. (1992). Geographical information science. *International Journal of Geographical Information Systems*, 6(1):31–45.

CHAPTER 1

Geographic Information Science: Defining the Field

David M. Mark
Department of Geography, University of Buffalo
Buffalo, NY, 14261, USA

1 INTRODUCTION

In the last decade, geographic information science has emerged as a focus of considerable academic attention. To some extent, it is the Earth's New Science, just as cognitive science was the Mind's New Science a decade or two earlier (Gardner, 1985). But it is not clear how deep or lasting the impact of GI science will be, either on academia or on the GIS industry. Rather than following the success of cognitive science, GI science could just as easily be the next regional science, a similar fusion of disciplines that peaked early and continues today mainly as an internally successful multidisciplinary field of relatively low influence on science, technology, or society. Worse yet, geographic information science could largely be just a pretentious name for geographic information systems, and not really a scientific or intellectual field at all. This paper seeks to explore these issues, and to lay out the intellectual scope of geographic information science.

1.1 What is geographic information science?

What is geographic information science? In the article in which he introduced the term, Goodchild (1992) did not provide a definition, but rather outlined the scope of the field indirectly by listing the major components of the geographic information science research agenda.

A written definition of the field followed when, in December 1994, a group of academics met in Boulder, Colorado, to establish a new organization to represent the GIS basic research committee. There was much debate over the name of the nascent organization, and votes over each word in the title: they settled on calling it the University Consortium for Geographic Information Science (UCGIS) (Mark and Bossler, 1995). Having chosen that name, the group was more or less compelled to provide a definition for the field. Again, though, the definition they provided was somewhat indirect:

> "The University Consortium for Geographic Information Science is dedicated to the development and use of theories, methods, technology, and data

for understanding geographic processes, relationships, and patterns. The transformation of geographic data into useful information is central to geographic information science." (UCGIS, 2002)

A full definition of GI science was provided in a report on a workshop held in January 1999 at the National Science Foundation, Geographic Information Science:

> "Geographic Information Science (GIScience) is the basic research field that seeks to redefine geographic concepts and their use in the context of geographic information systems. GIScience also examines the impacts of GIS on individuals and society, and the influences of society on GIS. GIScience re-examines some of the most fundamental themes in traditional spatially oriented fields such as geography, cartography, and geodesy, while incorporating more recent developments in cognitive and information science. It also overlaps with and draws from more specialized research fields such as computer science, statistics, mathematics, and psychology, and contributes to progress in those fields. It supports research in political science and anthropology, and draws on those fields in studies of geographic information and society." (Mark, 2000)

However, the community has not fully adopted such a definition of GI science as a fundamental research field. Recently, at least in the United States, there has been a tendency to use the term "geographic information science" to refer to almost any academic work that involves geographic information systems (GIS), often without changing the content of conventional "GIS" teaching programs at all. Despite this, it appears that the field has considerable depth, and a richness of intellectual challenges that mark it as a legitimate multidisciplinary field and perhaps an emerging new discipline. Thus, the author is comfortable asserting that GI science is most certainly not just a new name for GIS training and applications. In a purist view of GI science, even the use of GIS as a tool in scientific research is not GI science—at a recent NSF workshop, this latter area was termed "Research Using GIS" (Mark, 2000). Research Using GIS, is important to the sciences, and to funding for GIS-related scientific work, but is not GI science *per se*. Of course, since current commercial GIS may not have all the capabilities to support scientific research in other fields, Research Using GIS may reveal important research topics for GI science to address, and contribute to the GI science research agenda.

1.2   What are information and information science?

Given the name chosen for the field, it would appear that geographic information science would be a branch of information science. Does adding the word "geographic" to a good definition of information science produce a reasonable definition of GI science? Unfortunately, it is difficult to find good definitions of information. The Oxford English Dictionary provides several definitions. The definitions that appear closest to what we are dealing with in GI science are:

> 3a. "Knowledge communicated concerning some particular fact, subject, or event; that of which one is apprised or told; intelligence, news. spec. contrasted with data."

3c. "that which inheres in one of two or more alternative sequences, arrangements, etc., that produce different responses in something, and which is capable of being stored in, transferred by, and communicated to inanimate things."

Next, "information science" is defined in the OED as follows:

"(that branch of knowledge which is concerned with) the procedures by which information, esp. that relating to technical or scientific subjects, is stored, retrieved, and disseminated."

Although the life of philosopher Edmund Husserl (1859–1938) pre-dates the computer age, he provided an intriguing way to characterize information (Smith, 1989). When a mental act of an intelligent agent is directed towards an object in the real world, information is produced. In this view, information is a key part of the content of mental acts, specifically the conceptual or communicable part of the content of mental acts. This definition of information ties it to conceptualisations, intelligent agents, and cognition, which may prove especially valuable.

A practical definition of information science is provided by Shuman. After confirming the impression that "information science is very difficult to define" (p. 15), Shuman states:

"Information science is very difficult to define. ... the field of information science, however, may be defined as one that investigates the properties and behavior of information, how it is transferred from one mind to another, and optimal means for making that transfer, in both natural and artificial systems. Finally, information science is concerned with the effects of information on people and on machines." (Shuman, 1992, p. 15)

Insert the word "geographic" in front of "information," and this appears to be a reasonable definition for our field, and the contention that geographic information science is, fundamentally, a branch of information science, seems quite tenable.

2 HISTORY OF GEOGRAPHIC INFORMATION SCIENCE

The history of GI Systems in itself an important topic worthy of much study, and the article by Coppock and Rhind (1991) and the edited book by Foresman (1998) are excellent beginnings that only scratch the surface. Recall, however, that this chapter is about geographic information *science*, not about geographic information *systems*, and thus restricts the history discussion to the emergence of the science.

2.1 The NCGIA solicitation

Although geographic information systems and computer cartography generated a number of challenging conceptual and computational problems from their initiation in the 1960s or even earlier, with academics involved since the early days, the idea that there might be an academic field of study, a science, behind GIS software technology, came in the 1980s. A pivotal event in the development of the science came when the field was targeted for significant funding by the US National Science Foundation (NSF). In 1984, Ron Abler joined the NSF as Program Director for the Geography and Regional Science program. Soon, he saw an opportunity to obtain NSF support for a science and technology research centre specifically devoted to GIS, as an opportunity to direct some of NSF's interest in research centres and 'Big Science' toward the discipline of geography. Abler wrote the NSF solicitation for a National Center for Geographic Information and Analysis (NCGIA) after meeting with members of the GIS and quantitative geography research communities (Abler, 1987, p. 304). The scientific core of this solicitation was a set of five research topics that an NCGIA should consider addressing. Since these topics were set out as bullet points in the NSF solicitation, they later became known as "the five bullets":

- spatial analysis and spatial statistics;
- spatial relations and database structures;
- artificial intelligence and expert systems;
- visualization;
- social, economic, and institutional issues

In historical perspective, these five points might now be seen as the first definition of the scope of an emerging new research field that later became known as geographic information science. The successful proposal for NCGIA claimed that the third and fourth bullets, artificial intelligence and visualization, were crosscutting methodological themes, and emphasized three basic areas indicated by the first, second, and fifth bullets NCGIA (1989).

2.2 The naming of the field

The phrase "geographic information and analysis" is a grammatically awkward yet clever effort to integrate elements of GIS with the spatial analysis tradition from geography. At the same time, academic researchers in GIS were conducting active debates about the intellectual status of the GIS field. In July of 1990, Michael F. Goodchild made a keynote address about GIS-related research priorities to the Spatial Data Handling conference in Zurich, Switzerland, the fourth conference under that title. Goodchild's talk was entitled "Spatial Information Science." The address was subsequently published in the *International Journal of Geographical Information Science* (IJGIS) under the modified title "Geographical Information Science" (Goodchild, 1992); the shift from "Spatial" to "Geographical" in the title was not discussed. In the IJGIS article, Goodchild did not provide a crisp definition of the field, but did discuss what he called "the content of geographical information science," under the following eight headings:

1. Data collection and measurement
2. Data capture
3. Spatial statistics
4. Data modelling and theories of spatial data
5. Data structures, algorithms and processes
6. Display
7. Analytical tools
8. Institutional, managerial and ethical issues

2.3 The University Consortium for Geographic Information S...

During the 1980s, many sectors of the GIS community formed organizations to promote their interests (Brown et al., 2002). Academic researchers, while among the last, were not immune to this trend, and a series of discussions through 1992 and 1993 led to a founding meeting in Boulder, Colorado, in December 1994. In June 1996, UCGIS delegates met to determine research priorities for the new organization, and at the end of the meeting endorsed 10 research priorities.

- Spatial Data Acquisition and Integration
- Distributed Computing
- Extensions to Geographic Representation
- Cognition of Geographic Information
- Interoperability of Geographic Information
- Scale
- Spatial Analysis in a GIS Environment
- The Future of the Spatial Information Infrastructure
- Uncertainty in Spatial Data and GIS-based Analyses
- GIS and Society

There are four topics on the UCGIS research challenges that were not on Goodchild's list of topics. One highly applied topic, "The Future of the Spatial Information Infrastructure," is subsumed under "GIS and Society" in the other schemes. The other three are more basic: interoperability, distributed computing, and scale. Of these, scale is a compelling, crosscutting theoretical and conceptual issue, whereas interoperability and distributed computing appear to be cross-cutting technological themes.

2.4 Project Varenius

In 1995, researchers from the NCGIA submitted a proposal entitled "Advancing Geographic Information Science" to the NSF. This proposal defined a new vision of GI science as a field based on three fundamental research areas:

- Cognitive Models of Geographical Space
- Computational Methods for Representing Geographical Concepts
- Geographies of the Information Society

Seen in the light of the agendas compared in this paper, the Varenius triangle is rather unbalanced, under-playing the computational components of the field. Looking at the research specifics of the Varenius project, the computational component addressed interoperability, ontology of fields, and geographic knowledge discovery and data mining.

3 COMPONENTS OF GEOGRAPHIC INFORMATION SCIENCE

The nature of the field of geographic information science can be characterized by listing its components. These topics are not meant to define a research agenda. It is more like a curriculum for the topic, the basic components of the field even if research on them is relatively mature or complete. The topics are interlinked, and there is no single linear sequence that would fully preserve their relationships. Thus, the sequence employed below is somewhat arbitrary. Lastly, the headings are neither definitive nor exhaustive, and several important topics that do not readily fit into the scheme are reported under the heading "Other geocomputation topics" below.

3.1 Ontology and representation

1. Ontology of the geographic domain

Ontology deals with what exists, and with what may possibly exist. In this sense, it is a branch of philosophy that deals with some of the most fundamental aspects of scientific inquiry, but at a very high level of abstraction. This part of GI science examines the geographic information and geographic concepts that are used by environmental and social scientists in their research, as well as by people in general. Ontology seeks to provide a consistent formal theory of tokens (instances) and types (kinds) in the real world, their relationships, and the processes that modify them. Overviews of this topic are provided in Smith and Mark (1998, 2001). Recently, ontology was identified as an emerging theme by UCGIS, and was the topic for a special issue of the IJGIS (Winter, 2001).

2. Formal representation of geographic phenomena

More recently, the term ontology has been used in information science and knowledge representation to refer to the specifications of the conceptualisations employed by different groups of users in regard to domains of entities of different types. Characteristically, such specification involves the laying down of a computationally tractable taxonomy of the objects in the given domain of a sort that can support automatic translation from one data context to another. These representations are types or kinds in the digital domain, to be instantiated through data to become digital tokens (instances) that correspond to geographic things in reality. This topic, finding digital formalisms that can capture the essence of geographic phenomena, has traditionally been referred to in GIS as data modelling or as representation, but can also be seen as the applied side of ontology. The DIME encoding system for street networks and census districts (Cooke and Maxfield, 1967) may have been the greatest ever innovation in geographic representation, but there have been many others, and there is certainly still a need for improvements in this area (Yuan et al., in press).

3.2 Computation

3. Qualitative spatial reasoning

Reasoning about spatial relations and positions in space is a well-established research area in artificial intelligence, and has become important in GI science as well. Cognitive and linguistic models of spatial relations predominantly involve qualitative topological principles such as contact and containment. There are two domains for spatial relations. When spatial objects are disjoint, their spatial relation is characterized by distance and direction. Distance and direction can be reported quantitatively as metric distances and angles, but qualitative models of distance and direction correspond well with human reasoning and are appropriate for many purposes (Frank, 1992). The other domain for spatial relations is when the objects touch or overlap. Recent work on relations between non-disjoint spatial objects has been dominated by two formal frameworks. Egenhofer and his colleagues have based their work on point sets (Egenhofer and Franzosa, 1991) and extended this under the 9-intersection formalism (Egenhofer et al., 1994). A competing framework is provided by the RCC family of formalisms by Cohn and his colleagues (Cohn et al., 1997; Cohn and Hazarika, 2001).

4. Computational geometry

Computational geometry provides fundamentals for metric representation of objects and relations in geographic space. Analytical cartography is an alternative term for many aspects of computational geometry applied to the geospatial domain. Computational solutions to geometric problems were required in the very earliest days of computer-assisted cartography and GIS. Computational geometry is challenging, in part because pure Euclidean geometry cannot be implemented in straightforward fashion in a finite-precision digital environment, as discussed in an early paper by Douglas (1974) and discussed in detail by Franklin (1984). Line simplification (Douglas and Peucker, 1973) and many other aspects of map generalization (Buttenfield and McMaster, 1991) fall under the general topic of computational geometry in GI science, although they also relate to the cross-cutting issue of scale (below). Another set of computational geometry problems relate to the efficient computation of proximity, handled under the conceptual framework variously labelled as Voronoi diagrams, or Thiessen or proximal polygons. Preparata and Shamos (1991) provided a definitive review of these problems, and Gold has done much work to integrate these methods into geographic information science (see Gold, 1994, for example).

5. Efficient indexing, retrieval, and search in geographic databases

Efficient indexing of multidimensional data is an important problem in database research in computer science. Since geographic information is inherently at least two dimensional, these indexing issues have long been important in GIS. The so-called Morton 'matrix' approach for ordering map areas on a sequential magnetic tape was a key early innovation in GIS (Morton, 1966). Morton's index was equivalent to interleaving the bits in x- and y-coordinates expressed as integers. The idea was re-discovered in the context of image processing and retrieval in the 1970s under the label quadtrees, which recursively divide an image into quadrants and subquadrants (see Samet, 1989, for a review). Samet (1989) reviews a number of related indexing schemes such as B-trees, R-trees, k-d trees, etc.

6. Spatial statistics

Spatial statistics is an important research area with strong links to geographic information science. One of the properties that make spatial information special is the frequent presence of spatial autocorrelation or spatial dependence. Spatial statistics (Cressie, 1993) provides formal statistical methods for dealing with spatial autocorrelation, such as measuring it, or controlling for its effects when conducting statistical analyses based on data for spatial units. Spatial statistics can be used to characterize some aspects of data quality, but otherwise this topic appears to stand in some isolation from other components of geographic information science.

7. Other geocomputation topics

A number of additional computational topics are important to GI science but do not fit under the headings that immediately precede this one. One of these topics is map algebra, a comprehensive conceptual framework for raster-based spatial analysis developed in several articles and summarized in a 1990 book (Tomlin, 1990). This is not just a matter of implementing standard GIS operations based on a different representation of spatial information. Rather, map algebra leads to a different way of conceptualising geocomputational problems based on proximity and local operators that could easily re-cast in a parallel computation environment. Closely related to map algebra are the many spatial operations that can be based on cellular automata (von Neumann, 1966; Couclelis, 1997; Egenhofer et al., 1999).

3.3 Cognition

8. Cognitive Models of Geographic Phenomena

This research area involves the study of human perception, learning, memory, reasoning, and communication of and about geographic phenomena. An explicit agenda to examine human cognition of geographic environments was originally introduced into the GI science agenda as a way to gain insight into the nature of spatial relations, to gain insights into geographic ontology, and to understand and improve human-computer interaction for GIS (Mark and Frank, 1991). There is a large body of work on spatial cognition in psychology and cognitive science, and some of this has dealt with the geographic domain— some benchmark examples include Stevens and Coupe (1978), Talmy (1983) and Herskovits (1986). Studies of human spatial cognition are foundational to several other areas of GI science. Attention to formalizing common-sense concepts for geographic space was highlighted by Egenhofer and Mark (1995) under the term 'naïve geography'.

9. Human interaction with geographic information and technology

Human-computer interaction (HCI) for geographic information systems, and the design of user interfaces for GIS, is perhaps the most obvious example of the relevance of cognition to GIS (Mark and Gould, 1991; Medyckyj-Scott and Hearnshaw, 1993; Nyerges et al., 1995). The importance of this topic as a part of the GI science research agenda depends on whether the issues of GIS usability can be separated into general issues of human-computer interaction on the one hand and issues of geographic concepts on the other. If not, then the GI science research community must address problems in the overlap.

3.4 Applications, institutions, and society

10. Acquisition of geographic data

For all its attention to theory, information, knowledge, and wisdom, data or measurements of positions and attributes of the geographic domain is still central to GIS. This component of GI science starts with a solid theory of measurement in general, and builds this out to provide an account of how the geographic world and its conceptualisations are measured and converted to instantiations of the aforementioned geographic representations. If narrowly defined, geomatics is a good term for this component of geographic information science. Technologies for the acquisition of geographic data and information, chiefly remote sensing and GPS, are fields in their own rights that dwarf the rest of GI science in terms of the public and private investments in their infrastructures.

11. Quality of geographic information

Research on the quality of geographic information, including variations under the terms accuracy or error, are important parts of the GI science research agenda (Goodchild and Gopal, 1989). Data quality may be reduced due to measurement error during data acquisition or by various transformations of the data in GIS processing. Specification errors may occur if the ontology or representation is not done correctly. The impacts of data quality must be judged in terms of the sensitivity of models and the fitness of data for particular uses.

12. Spatial analysis

Spatial analysis is an important topic in quantitative geography, with strong ties to many aspects of geographic information science. Some might argue that spatial analysis is just an application of GI science principles to problems in environmental or social science, and that the fundamental science underlying spatial analysis is already covered by the ontology, representation, and spatial statistics topics. However, several topics, especially the "modifiable areal unit problem" (MAUP; Openshaw and Taylor, 1981; Openshaw, 1984) seems clearly to be an important class of GI science research and curriculum topics, yet is not covered well in other parts of the GI science agenda. Careful deliberation will be needed to divide spatial analysis into topics that are part of GI science, and topics that are applications of GI science.

13. Geographic information, institutions, and society

Societal impacts of GIS technology has been a part of the real agenda of GIS since its onset—indeed, before 1980, most technical innovation in GIS occurred in a direct application context, as software was developed at government agencies or by consultants, rather than by academics. The 1980s were a decade of commercialisation of software and data on the one hand and 'academisation' of the GIS research and development agenda on the other. Since the end of the 1980s, academic attention returned to the impacts of the technology from a new direction—post-Modern critics of quantification and technology. At the same time, there was also an increase in research on economic and legal aspects of geographic information, its production and sharing, including studies of how the use of GIS and associated technologies by individuals and institutions changes efficiency, effectiveness, equity, and power in society. A particularly active area of research and practice is "Public-Participation GIS" (PPGIS). The main aspects of the post-Modern critique have been presented by Pickles in his edited book (Pickles, 1995) and a more recent review article (Pickles, 1999). A good overview of the economic and legal aspects of geographic information is found in Masser and Onsrud (1993).

3.5 Crosscutting research themes

14. Time

Time and temporality, motion and change, are essential to many GIS applications, yet GIS software has been notoriously weak in providing tools for handling temporal dimensions of geographic information. Long ago, Blaut (1961) proposed that the Kant/Newton separation of reality into space, time, and theme, which Berry (1964) proposed as an organizing framework for geography and GIS, made it difficult to deal with process and change. GIS seems ontologically committed to separating space and time, which then impedes certain scientific uses of GIS. If space and time can truly be studied separately and the results later assembled, then time would not be part of the agenda for GI science. Recent interest in time in geographic space and GIS (Langran, 1992; Peuquet, 1994; Egenhofer and Golledge, 1998) suggests that time is an integral part of GI science research and one which cuts across most other GI science topics.

15. Scale

Scale has multiple meanings relevant to GI science. In cartography, scale refers to size on the map divided by size in the world—small-scale maps show large regions. Map scale interacts with geometry of the world to require map generalization (Buttenfield and McMaster, 1991). In the physical sciences such as meteorology or geomorphology, the term scale is used to indicate the size, extent, or characteristic length for physical processes. Micro-, meso-, and global-scale atmospheric processes are familiar terms. In biology and geomorphology, interactions between size, shape, and function have been variously explored under the heading of allometry—sometimes shape must change systematically in order to maintain function as size changes (for a review, see Church and Mark, 1980). Thirdly, the term scale is used to summarize resolution, the smallest entities that can be detected or represented, both for display and analysis (see Quattrochi, 1997). More recently, cognitive aspects of scale have been highlighted (Montello, 1993; Montello and Golledge, 1999). The issues surrounding the term scale are important to the GI science curriculum and to the research agenda, and cut across most of the other topics described in this paper.

4   COMPARISON OF TOPICS

Of course, the list of components of geographic information science proposed in this paper was prepared in full knowledge of the lists proposed by Goodchild (1992) and by the UCGIS. Still, it is instructive to compare the lists.

Topics on all three lists

Four topics appear on all three lists:

- data acquisition
- representation (data modelling)
- spatial analysis
- societal issues surrounding GI

The last three of these also correspond well with core issues in the bullets proposed in NSF's NCGIA solicitation.

Topics on two lists

Three topics from Goodchild's (1992) list also were identified as key topics in this paper, but were not on the UCGIS list:

- computational geometry
- indexing for spatial databases
- spatial statistics

In fact, Goodchild combined the first two of these topics under his " Data structures, algorithms and processes."
Three additional topics were on the UCGIS list and in this paper but not singled out by Goodchild:

- ontology
- cognition
- data quality

Another research topic, scale, is a research topic on the UCGIS list but identified as a crosscutting research theme here.
Lastly, one topic, visualization and display, was on Goodchild's (1992) list, and recently was identified as an emerging research theme by UCGIS, but is not on the list of GI science fundamentals described in this paper; perhaps it should join scale and time as cross-cutting themes.

Topics on only one list

All of Goodchild's topics from 1992 appear on one or both of the other lists. However, three of UCGIS's original research challenges, and one of their emerging themes, appear on neither of the other lists:

- distributed computing
- interoperability of geographic information
- the future of the spatial information infrastructure
- geospatial data mining and knowledge discovery

Proposed in this paper, but not on either of the other lists, are four additional topics:

- human interaction with GI and technology
- qualitative spatial reasoning
- time in geographic space
- other geocomputation topics

The human-computer interaction topic is partially covered by UCGIS under its version of the "Cognition" topic, and dealing with time is part of the material in UCGIS topic, "Extensions to Geographic Representations."

5 DISCUSSION

How can we put these proposed components of geographic information science in perspective? One rather narrow but revealing way to look at the prospects for progress in GI science is to compare the research priorities outlined above to the priorities of funding agencies. The programs of the Information and Intelligent Systems (IIS) Division of the Computer and Information Science and Engineering (CISE) Directorate at the US NSF are very interesting in this regard. As of the summer of 2000, NSF's IIS Division was divided into 8 programs:

- Computation and Social Systems
- Human Computer Interaction
- Information and Data Management
- Information Technology Research
- Knowledge and Cognitive Systems
- Robotics and Human Augmentation
- Special Projects (IIS)
- Digital Libraries

The similarity of this subdivision of information science aspects of computer science research to the GI science agenda is striking. This provides good evidence regarding the claim that geographic information science should be regarded as a branch of information science. Of course, geographic information science has an intimate relationship with the discipline of geography, since they address the same aspects of reality. However, geographic information science is concerned with ontology, representation, and computational issues, whereas geography attempts to explain and predict geographic phenomena.

Material presented in this paper confirms that geographic information sscience has considerable depth, and a richness of intellectual challenges that mark it as a legitimate multidisciplinary field and perhaps an emerging new discipline. In the writer's opinion, progress in the field would be aided by a consensus among leading researchers on the nature, scope, and elements of the field; this paper attempts a step in that direction.

REFERENCES

Abler, R. F. (1987). The National Science Foundation National Center for Geographic Information and Analysis. *International Journal of Geographical Information Systems*, 1(4):303–326.

Berry, B. J. L. (1964). Approaches to regional analysis: A synthesis. *Annals of the Association of American Geographers*, 54:2–11.

Blaut, J. M. (1961). Space and process. *The Professional Geographer*, 13:1–7.

Brown, D. G., Elmes, G., Kemp, K. K., Macey, S., and Mark, D. (2002). Geographic information systems and science. In Gaile, G. and Willmott, C. (Eds), *Geography In America 2002*. In press.

Buttenfield, B. P. and McMaster, R. B. (Eds) (1991). *Map Generalization: Making Decisions for Knowledge Representation*. London: Longmans Publishers.

Church, M. and Mark, D. M. (1980). On size and scale in geomorphology. *Progress in Physical Geography*, 4:342–390.

Cohn, A. G., Bennett, B., Gooday, J. M., and Gotts, N. (1997). Qualitative spatial representation and reasoning with the region connection calculus. *GeoInformatica*, 1:275–316.

Cohn, A. G. and Hazarika, S. M. (2001). Qualitative spatial representation and reasoning: An overview. *Fundamenta Informaticae*, 46(1–2):1–29.

Cooke, D. F. and Maxfield, W. H. (1967). The development of a geographic base file and its uses for mapping. In *Proceedings of the Fifth Annual URISA Conference*, pp. 207–218.

Coppock, J. T. and Rhind, D. W. (1991). The history of GIS. In Maguire, D. J., Goodchild, M. F., and Rhind, D. W. (Eds), *Geographical Information Systems: Principles and Applications*, volume 1, pp. 21–43. London: Longmans Publishers.

Couclelis, H. (1997). From cellular automata to urban models: New principles for model development and implementation. *Environment and Planning B: Planning & Design*, 24:165–174.

Cressie, N. (1993). *Statistics for Spatial Data*. New York: J. Wiley.

Douglas, D. H. (1974). It makes me so cross. Laboratory for Computer Graphics and Spatial Analysis, Graduate School of Design, Harvard University. (Reprinted in Marble, D., Calkins, H. and Peuquet, D., Eds, 1984. *Basic Readings in Geographic Information Systems*, Williamsville NY: SPAD Systems Ltd.).

Douglas, D. H. and Peucker, T. K. (1973). Algorithms for the reduction of the number of points required to represent a digitized line or its caricature. *The Canadian Cartographer*, 10(2):112–122.

Egenhofer, M. F., Glasgow, J., Gunther, O., Herring, J., and Peuquet, D. (1999). Progress in computational methods for representing geographic concepts. *International Journal of Geographical Information Science*, 13(8):775–796.

Egenhofer, M. F., Mark, D. M., and Herring, J. (1994). The 9-intersection: Formalism and its use for natural-language spatial predicates. Technical Report 94-1, National Center for Geographic Information and Analysis, Santa Barbara, CA.

Egenhofer, M. J. and Franzosa, R. D. (1991). Point-set spatial relations. *International Journal of Geographical Information Systems*, 5:161–174.

Egenhofer, M. J. and Golledge, R. (Eds) (1998). *Spatio-Temporal Reasoning in Geographical Information Systems*. London: Oxford University Press.

Egenhofer, M. J. and Mark, D. M. (1995). Naïve geography. In Frank, A. U. and Kuhn, W. (Eds), *Spatial Information Theory: A Theoretical Basis for GIS*, volume 988 of *Lecture Notes in Computer Science*, pp. 1–15. Berlin: Springer-Verlag.

Foresman, T. W. (Ed) (1998). *The History of Geographic Information Systems: Perspectives from the Pioneers*. Prentice-Hall.

Frank, A. U. (1992). Qualitative spatial reasoning about distances and directions in geographic space. *Journal of Visual Languages and Computing*, 3:343–371.

Franklin, W. R. (1984). Cartographic errors symptomatic of underlying algebra problems. In *Proceedings, International Symposium on Spatial Data Handling*, pp. 190–208, Zurich, Switzerland.

Gardner, H. (1985). *The Mind's New Science: A History of the Cognitive Revolution*. New York: Basic Books.

Gold, C. M. (1994). Three approaches to automated topology, and how computational geometry helps. In *Proceedings, Sixth International Symposium on Spatial Data Handling: Advances in GIS Research*, pp. 145–158, Edinburgh, Scotland.

Goodchild, M. and Gopal, S. (Eds) (1989). *The Accuracy of spatial databases*. London: Taylor & Francis.

Goodchild, M. F. (1990). Spatial information science. In *Proceedings, Fourth International Symposium on Spatial Data Handling*, pp. 3–12, Zurich, Switzerland.

Goodchild, M. F. (1992). Geographical information science. *International Journal of Geographical Information Systems*, 6(1):31–45.

Herskovits, A. (1986). *Language and Spatial Cognition: A Interdisciplinary Study of the Prepositions in English*. Cambridge, England: Cambridge University Press.

Langran, G. (1992). *Time in Geographic Information Systems*. London: Taylor & Francis.

Mark, D. M. (2000). Geographic information science: Critical issues in an emerging cross-disciplinary research domain. *Journal of the Urban and Regional Information Systems Association*, 12(1):45–54.

Mark, D. M. and Bossler, J. (1995). The university consortium for geographic information science. *Geo Info Systems*, 5(4):38–39.

Mark, D. M. and Frank, A. U. (Eds) (1991). *Cognitive and Linguistic Aspects of Geographic Space*. Dordrecht: Kluwer Academic Publishers.

Mark, D. M. and Gould, M. D. (1991). Interacting with geographic information: A commentary. *Photogrammetric Engineering and Remote Sensing*, 57(11):1427–1430.

Masser, I. and Onsrud, H. (Eds) (1993). *Diffusion and Use of Geographic Information Technologies*. Dordrecht, Netherlands: Kluwer Academic Publishers.

Medyckyj-Scott, D. and Hearnshaw, H. M. (Eds) (1993). *Human Factors in Geographical Information Systems*. Belhaven Press.

Montello, D. (1993). Scale and multiple psychologies of space. In Frank, A. U. and Campari, I. (Eds), *Spatial Information Theory, A Theoretical Basis for GIS*, volume 716 of *Lecture Notes in Computer Science*, pp. 312–321. Berlin: Springer-Verlag.

Montello, D. R. and Golledge, R. G. (1999). Scale and detail in the cognition of geographic information. Technical Report Specialist Meeting, Project Varenius, University of California at Santa Barbara, Santa Barbara, CA. http://www.ncgia.org.

Morton, G. (1966). A computer oriented geodetic database and new technique in file sequencing. IBM Canada.

NCGIA (1989). The research plan of the National Center for Geographic Information and Analysis. *International Journal of Geographical Information Systems*, 3(2):117–136.

Nyerges, T. L., Mark, D. M., Laurini, R., and Egenhofer, M. F. (Eds) (1995). *Cognitive Aspects of Human-Computer Interaction for Geographic Information Systems.* Dordrecht, Netherlands: Kluwer Academic Publishers.

Openshaw, S. (1984). *The Modifiable Areal Unit Problem.* Number 38 in Concepts and Techniques in Modern Geography. Norwich: Geo Books.

Openshaw, S. and Taylor, P. J. (1981). The modifiable areal unit problem. In Wrigley, N. and Bennett, R. J. (Eds), *Quantitative Geography: A British View.* London: Routledge.

Peuquet, D. J. (1994). It's about time: A conceptual framework for the representation of temporal dynamics in geographic information systems. *Annals of the Association of American Geographers*, 84(3):441–461.

Pickles, J. (Ed) (1995). *Ground Truth: The Social Implications of Geographic Information Systems.* New York: Guilford Press.

Pickles, J. (1999). Arguments, debates and dialogues: The GIS-social theory debate and the concern for alternatives. In Maguire, D. J., Goodchild, M. F., Rhind, D. W., and Longley, P. (Eds), *Geographical Information Systems: Principles and Applications*, volume 1, pp. 49–60. New York: John Wiley & Sons, 2nd edition.

Preparata, F. and Shamos, M. E. (1991). *Computational Geometry: An Introduction.* Texts and Monographs in Computer Science. Berlin: Springer-Verlag.

Quattrochi, D. A. (Ed) (1997). *Scale in Remote Sensing and GIS.* New York: CRC Press, Lewis Publishers.

Samet, H. (1989). *The Design and Analysis of Spatial Data Structures.* Reading, MA: Addison-Wesley.

Shuman, B. A. (1992). *Foundations and Issues in Library and Information Science.* Englewood, Colorado: Libraries United Inc.

Smith, B. (1989). Logic and formal ontology. In Mohanty, J. N. and McKenna, W. (Eds), *Husserl's Phenomenology: A Textbook*, pp. 29–67. Lanham: University Press of America.

Smith, B. and Mark, D. M. (1998). Ontology and geographic kinds. In Poiker, T. K. and Chrisman, N. (Eds), *Proceedings, Eighth International Symposium on Spatial Data Handling (SDH'98)*, pp. 308–320, Vancouver. International Geographical Union.

Smith, B. and Mark, D. M. (2001). Geographical categories: An ontological investigation. *International Journal of Geographical Information Science*, 15(7):591–612.

Stevens, A. and Coupe, P. (1978). Distortions in judged spatial relations. *Cognitive Psychology*, 10:422–437.

Talmy, L. (1983). How language structures space. In Pick, H. L. and Acredolo, L. P. (Eds), *Spatial Orientation: Theory, Research, and Application*, pp. 225–282. New York: Plenum Press.

Tomlin, C. D. (1990). *Geographic Information Systems and Cartographic Modeling.* Englewood Cliffs, NJ: Prentice-Hall.

UCGIS (2002). UCGIS bylaws. http://www.ucgis.org/fByLaws.html. Quote from 14 June 2002 version.

von Neumann, J. (1966). *Theory of Self-Reproducing Automata*. Champain, IL: University of Illionois Press.

Winter, S. (2001). Ontology: Buzzword or paradigm shift in GI science? *International Journal of Geographical Information Science*, 15(7):587–590.

Yuan, M., Egenhofer, M., Mark, D. M., and Peuquet, D. J. (in press). Extensions to geographic representations. In McMaster, R. B. and Usery, L. (Eds), *Research Challenges in Geographic Information Science*. New York: John Wiley & Sons.

CHAPTER 2

The Nature and Value of Geographic Information

Michael F. Goodchild
Department of Geography, University of California
Santa Barbara, CA, 93106, USA

1 INTRODUCTION

Markets for commodities such as wheat, crude oil, or pork bellies depend on the ability to measure quantity, whether it be in bushels, barrels, or tons. On the other hand information-rich commodities such as books or newspapers are priced per unit rather than by such measures as volume (despite apocryphal stories of assessing dissertations by weight), since they are always bought and sold whole, and there is little point in trying to establish a price for half a book, or a fraction of a newspaper. Similarly paper maps are priced by unit, rather than by square kilometer or some other continuous-scaled measure.

One of the underlying principles of the information economy is that digital information can also be traded as a commodity. But for some types of information the transition to digital form undermines the concept of a unit, and with it the information's basic granularity. Geographic information is a case in point, since digital geographic information is often merged into seamless databases, and disseminated for user-specified areas. A user might be interested in purchasing only the data covering a circular area surrounding a point, or only the data covering an irregularly shaped county or school district. In such cases a market for geographic information requires well-defined measures of information quantity, because otherwise all transactions would be unique, and no market could exist. For example, it must be possible to establish the combined value of two data sets given the values of each independently, using a rule such as a simple summation that obeys formal mathematical principles.

Buyers and sellers of digital geographic information have struggled with this issue, but little progress appears to have been made. The research initiative of the US National Center for Geographic Information and Analysis on the Use and Value of Geographic Information discussed this and many related issues (Onsrud and Calkins, 1993) without reaching definite conclusions, and other more recent reviews have similarly failed to report substantive progress (e.g. Longley et al., 2001). Instead, the community has adopted such measures as data volume (in bytes), and geographic area of coverage. But it makes little sense to price information by volume if such measures are sensitive to simple manipulation. For example, suppose a 15MB data set is compressed to 3MB using a loss-less compression such as run-length encoding; is it now worth only 20% of its previous value? What if a vector data set is restructured in raster form, and as a result grows in volume by a factor of ten? Moreover, pricing by area of coverage makes little sense if the volume of information, and the cost of collecting it, varies dramatically from one area to another.

A satisfactory measure of information quantity would have to allow price to be determined for any user-defined area. It would have to be independent of data structure, since simple restructuring (e.g., from one data format standard to another) should not affect price. It would have to be independent of medium, if price is to be determined by data content rather than by the medium on which the content is stored. Methods should also be available to determine if two data sets have the same content, and differ only by such comparatively irrelevant factors as medium, or data structure, or projection. One of the advantages of digital information is the ease with which it can be mutated into other forms, through simple transformations. A transformation is said to be *reversible* if its inputs can be recovered by a simple reverse transformation. Today's geographic information systems make it easy to perform a range of reversible transformations, including changes of projection and datum, or topological overlay. A satisfactory measure of information quantity clearly should not be affected by reversible transformation (see Kuhn, 1997, for a comprehensive discussion of this issue).

These requirements can only be achieved through a theory of geographic information that is *semantic* rather than *syntactic*; in other words, one that focuses on the meaning of information rather than on its form. A semantic measure of information should also distinguish between information that adds to the receiver's knowledge, and information that merely duplicates existing knowledge; in the second case the information content should be zero. A semantic measure of information must therefore depend not only on the content of the message, but also on various aspects of the state of the receiver's knowledge, including knowledge of relevant rules and conventions.

This chapter is intended to make a few small steps in the direction of such a semantic measure, and associated theory. It begins with a discussion of classical syntactic information theory, in order to demonstrate its inadequacy. A theory is proposed in which geographic information is constructed from simple atomic tuples, and the problems of dealing with the potentially infinite number of such tuples are addressed. The subsequent section addresses queries, and the role of geographic information systems (GIS), within this theoretical framework. This is followed by a discussion of two issues of direct relevance to the framework: digital Earth, and naïve geography. The framework is used to develop a semantic theory of communication that satisfies the requirements outlined above. Finally, the chapter concludes with a short discussion of unaddressed issues.

Many aspects of the approach are general, and might be usefully applied to any type of information. But geographic information seems a particularly well-defined type or subset of information, and as such provides an excellent base for speculating about the nature of information in general.

1.1   Shannon-Weaver information theory

Syntactic theories of information have been developed extensively over the past half century, largely in the context of coded communication. Shannon (1949) proposed that the information content of a coded message could be measured in terms of the probabilities of occurrence of each possible code. Suppose, for example, that a message is sent in binary code as a string of 0s and 1s, and that in such messages the digits 0 and 1 occur with roughly equal frequency. Then a single digit in the message can be regarded as resolving between two equally likely possibilities, and its information content can be measured accordingly using metrics of the form $- p_i \log p_i$, where p_i is the probability associated with code i. Logs to base 2 are convenient when the code is binary, and in this case the units of the measure are known as *bits*. The measure is designed to remain constant despite a recoding of the message into some other coding scheme.

It is easy to extend this approach to more complex codes. For example, consider a single letter of the Roman alphabet and a message in the English language. Each letter now resolves among 26 possibilities, but in plain English the probabilities associated with each letter range widely, since E occurs much more frequently than X, and consequently can be said to convey less information. Further complications arise when sequences of more than one letter are considered, since the information content of a U following a Q in English is almost nil (its conditional probability is almost 1).

Certain forms of geographic information might be amenable to such an approach to measuring information content. Suppose a map of land cover type is coded using a raster structure, and each cover type i is associated with a probability p_i. Then the information content of the entire raster might be measured by treating it as a linear series of codes in the alphabet defined by the set of cover types. But it would be difficult to capture the effects of conditional probabilities, since in a typical geographic context a pixel surrounded by pixels of type A is more likely to be type A than any other type, and thus to have less information content. The approach can be applied hierarchically, to determine the information content of maps at different scales (for reviews of the use of information statistics in geographic research see Marchand, 1972; Thomas, 1981). In this context the use of information statistics is directly comparable to decompositions based on variance (e.g. Csillag and Kabos, 1996) or to Fourier or wavelet decompositions. One might similarly attempt to apply this approach to printed maps, based on the probability that any one point on the map is covered by any one of the map's inks. But here the number of points, and hence the length of the message, is potentially infinite. Moreover, it would be wrong to assess the points comprising a letter of annotation (such as the black ink denoting the letter A) as a set of independent codes, since they can also be regarded as a single code resolving among the 26 letters of the Roman alphabet.

This last problem is well illustrated using another simple example. Consider an infinite series of digits beginning with 3.14159... As an infinite series of apparently random decimal digits it has infinite information content. But the same information might be communicated using a single letter π from the Greek alphabet. What appears to be an example of infinite data compression is of course enabled by the convention that uses that particular Greek letter to denote that particular infinite series, and if the convention is not known to the receiver of the message then the point is lost and the message has no value.

Shannon-Weaver information theory is concerned with the form and coding of messages, and clearly fails to satisfy the requirements laid out above for a semantic theory of information. Such a theory would deal effectively with the last example, and recognize the equivalence of, first, an infinite transmission of digits; and second, the transmission of a Greek letter coupled with the knowledge that the receiver understands the relevant convention. It might also recognize a third case: the transmission of a computer code or mathematical rule for determining the digits of π. In principle, all of these three possibilities should have the same information content in a semantic theory of information.

2 A THEORY OF GEOGRAPHIC INFORMATION

Several attempts have been made to develop general theories of geographic information. Berry (1964) described a *geographic matrix*, a general model in which geography is characterized by location, time, and attributes. The same idea underlies Sinton's conceptualization (Sinton, 1978), as well as numerous discussions of geographic data modeling (e.g Peuquet, 1994).

Goodchild et al. (1999) propose that geographic information can be defined by reference to an *atomic element* or tuple of the form $\langle x, z \rangle$ where x denotes some location in space-time, and z denotes some set of properties associated with that location, commonly termed attributes. Location is constrained to the Earth's surface and near-surface. In this framework, all geographic information is ultimately reducible to a set of such atoms. For example, a digital elevation model consists of a series of atoms denoting the elevation at a regularly spaced series of points; elevation is the property associated with those locations; and time is ignored because elevation is assumed to be a static property, tectonic and other processes notwithstanding.

No message would be of any value to a receiver who did not understand the conventions underlying the message. In the case of geographic information, these include the conventions associated with position on the Earth's surface, such as latitude and longitude, datums and projections; and those associated with time, and we term these *universal locators*. With such universally accepted standards it is reasonable to expect that a receiver would understand x in most cases. But there is far less universality associated with z. While the Celsius scale of temperature is universal in the scientific world, the general public are more likely to use such vague terms as *warm* or *cold* to describe the temperature of the atmosphere near the ground. Other commonly encountered attributes such as land cover class are far less standardized, and the potential exists for far more confusion over the meaning of messages.

The dimensions of space and time that define x are continuous, and an infinite number of locations therefore exist. Thus while information gathered at points can be represented by individual atoms, an infinite number of atoms is required to characterize variation over lines, areas, or volumes. For example, it would require an infinite number of atoms to characterize the spatial variation of elevation over a finite area, or to define the extent of the State of California.

Two principles work to address this issue, and to make it possible to characterize geographic variation in a finite number of atoms. The first is often known as Tobler's Law of Geography (1970): "All things are related, but nearby things are more related than distant things." The effect is often measured using indices of spatial dependence, such as the Moran or Geary indices of spatial autocorrelation (Cliff and Ord, 1981) or the variogram (Isaaks and Srivastava, 1989), that compare variation over different distances. For example, the variogram plots average or expected variance between pairs of observations against the distance between them, and is normally observed to be a monotonically increasing function. It is easy to see the validity and generality of Tobler's Law if one tries to imagine a world in which it is not true. Such a world would be impossible to describe or inhabit, since the full range of variation could be encountered over vanishingly small distances.

In practice, many geographic phenomena exhibit constant values, or variation that lies below some acceptable threshold, over finite neighborhoods. The property "State of California" is uniformly true of all of the points within the state's boundaries, and similar principles apply to the attributes relevant to land ownership. In the case of land cover, class is held to be approximately constant within areas mapped as uniform class. For properties measured on continuous scales, such as elevation, Tobler's Law implies relationships of the form $|z(x + \delta x) - z(x)| < \lambda$ for $\delta x < \tau$ where z is such a property, τ defines a neighborhood in space and time, and λ defines an acceptable threshold of difference.

In other cases, Tobler's Law is implemented in the form of a series of *interpolation rules* that allow atomic tuples to be inferred from surrounding ones. For example, if temperature is known at a series of weather stations, then temperatures are inferred at intervening locations through simple processes such as Kriging or inverse-distance weighting (Longley et al., 2001).

The second principle derives from the process of determining location x. Location on the Earth's surface can be determined by a variety of means, including surveying, the use of the Global Positioning System, or by reference to a paper map. In each case the result is subject to measurement error, because of the limitations of the measuring instruments. Moreover, difficulties in defining the exact shape of the Earth, the tendency of the reference poles to wobble, and persistent crustal movements all work to ensure that measurement errors will never be reduced to zero. In practice, therefore, it is sufficient to represent geographic variation over a finite rather than an infinite number of points.

Moreover, the spatial resolution required by any application, or obtainable from any data acquisition system, is also strictly limited. Suppose a spatial resolution of λ is sufficient for a given application, λ being measured in linear units. Assuming a spherical Earth of radius R, the number of atoms required to characterize the spatial variation in some phenomenon over the surface of the Earth is approximately $4\pi R^2/\lambda^2$. The same number would be appropriate if λ instead represented the inherent spatial accuracy of the measurement of position (see, for example, Güting and Schneider, 1995).

The number of atoms needed to characterize the geographic domain is also dependent on the complexity of z, and the number of distinct properties present at any location. In principle an infinite number of properties might be held to exist, but in practice the number is finite, and strong correlations exist between many properties.

2.1 Fields and discrete objects

Because of these principles, and their effect in limiting the number of atoms, it has been possible to construct satisfactory representations of many geographic phenomena. Two approaches are used, representing alternative conceptualizations of variation in space (Worboys, 1995). The first, or *field* perspective, conceives of geographic variation in terms of a number of variables that are functions of position, $z = f(x)$. This conceptualization fits many physical phenomena, including land surface elevation, and atmospheric temperature and pressure. From the previous section, we know that a finite number of atoms is in practice sufficient to characterize a field with adequate accuracy over a finite space-time domain, and six approaches are commonly used in the case of two spatial dimensions and no temporal variation, as shown in Figure 1.

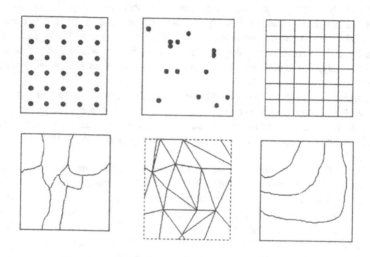

Figure 1: The six representations of a field commonly used in geographic databases. See text for explanation.

Row by row from the top left, the six approaches are as follows:

1. A finite number of regularly spaced points is used to create a finite number of atoms. Values at other locations are determined by an interpolation rule (for example, the rule that z is the value at the closest data point).
2. A finite number of irregularly spaced points is used. Values at intervening points are determined by an interpolation rule such as Kriging or inverse-distance weighting.
3. The area is partitioned into a finite number of rectangular regions, and a single value is recorded for each region. Detailed variation within regions is ignored.
4. The area is partitioned into a finite number of irregular polygonal regions, each defined by a number of points, and a single value is recorded for each region. Detailed variation within regions is ignored.

5. The area is partitioned into a finite number of irregular triangular regions. Atoms are recorded at the vertices, and values are interpolated within regions using linear functions (this approach can be generalized to include quadrilateral regions and nonlinear functions).
6. A finite number of isolines is defined. Each isoline is represented by an ordered set of points connected by straight segments, and a single value is recorded for each isoline (this approach can be generalized to allow for curved segments). Values between isolines must be obtained using an interpolation rule.

Each of these methods succeeds in characterizing the field, using a finite number of tuples. The form of the tuples varies; in (1), (2), and (5) the tuples are of the form $\langle x, z \rangle$, while in (4) and (6) they are of the form $\langle x_1, y_1, x_2, y_2, x_3, y_3, ..., x_n, y_n, z \rangle$, and in (3) a single tuple is sufficient, of the form $\langle z_1, z_2, z_3, ..., z_n \rangle$, given a known rule for ordering the elements of the raster, and a basis for its referencing to the Earth's surface.

In the *discrete object* perspective, the geographic world is conceived as empty, except where it is occupied by one or more points, lines, areas, or volumes. Each of these discrete objects has attributes describing its properties. Points are described by single atoms, and lines and areas by tuples of the form defined for cases (4) and (6) above in the static, two-dimensional case.

These methods succeed in creating data structures that use a finite number of tuples to describe geographic phenomena. Each can be decomposed into atomic tuples; for example, an infinite number of distinct tuples can be derived from any of the field representations and associated interpolation rules.

3 GEOGRAPHIC QUERIES

The previous section described a theory of geographic information based on distinct atoms or tuples. This section builds on that base by considering the nature of geographic queries, and the systems that respond to them. The term *geographic information system* (GIS) is used here to refer to such systems, and it is important to recognize that while most references to GIS imply a digital system, this same approach might be applied to the processes humans employ when responding to queries.

A *simple geographic query* is defined as a query to which an atomic tuple of geographic information is the answer. Two types of query exist, of the forms "What is at x?" and "Where is z?", and both are answered by providing the remaining part of the appropriate atomic tuple (or tuples). Two more advanced forms of query can be defined: those that can be answered from more complex structures without access to the atomic form, such as "What are the properties of this area?"; and those that require deduction from the contents of two or more atoms, such as "What is the distance from A to B?", which can be deduced from $\langle x_1, A \rangle$ and $\langle x_2, B \rangle$.

It is now possible to define what is meant by the possession of geographic information, in a manner that is independent of medium and structure. A GIS is said to possess an item of geographic information (defined as one or more atoms) if *it is capable of responding to a query to which the item is the answer*. This definition allows for the possibility that the GIS must undertake some form of transformation, processing, or deduction in order to obtain the answer. Note also that the definition defines possession in the context of a GIS rather than independently, implying that the system and the information are effectively inseparable.

Given such a definition, one might wonder whether it is possible to measure the quantity of information by counting the number of distinct queries that can be answered, and thus in effect counting atomic tuples. Consider, for example, the query "Is x in California?" A GIS could address this query by performing a simple point-in-polygon operation on x to determine whether the point lies inside a polygon representing the boundary of the State of California. But clearly an infinite number of such queries exist, since an infinite number of distinct points can be defined on the Earth's surface, and there exists an answer to the query for each of them. Does the GIS therefore possess an infinite amount of geographic information?

A previous section has already addressed this dilemma in a different form. If position is not knowable to better than some length λ, or if the location of the boundary of California is similarly known only to a limited accuracy, then the number of distinct queries is finite. Moreover, if x_1 is known to be in California and x_2 is also known to be in California, then there is a high probability that $\alpha x_1 + (1 - \alpha)x_2, 0 \leq \alpha \leq 1$ also lies in California (and a probability of 1 if California is convex), reducing the effective number of distinct queries further. Finally, the system is able to answer the query because it possesses a representation of the boundary of California as a polygon, plus an algorithm for resolving point-in-polygon queries. Thus the amount of information contained in the polygon definition could be said to place an upper bound on the amount of information possessed by the system, since the queries are resolved through a transformation of this information.

Another issue arises because of the possibility of answering queries corresponding to derivative information, rather than atomic tuples. For example, a system that *knows* $\langle x_1, A \rangle$ and $\langle x_2, B \rangle$ also is able to respond to a query about the distance between A and B. But this issue is easily addressed by restricting queries to atomic forms, and thus measuring information by the number of atoms of specifically geographic information.

3.1   Digital Earth

The phrase *Digital Earth* (DE) was coined by Gore (1992), and has been the focus of extensive discussion and the development of prototypes in recent years (see, for example, http://www.digitalearth.gov). DE is conceived as a repository of a vast array of information about the planet's surface and near-surface, in the future and past as well as the present, together with the means to render dynamic three-dimensional visualizations in virtual reality. In effect, it represents a single portal to all that is known about geography, that is, to all geographic information.

It is interesting to speculate on the nature of a complete DE. The previous discussion has shown that a finite number of atoms of geographic information can be sufficient to create a satisfactory representation, and Goodchild (2001) has argued that DE is technically feasible within the limits of today's Internet bandwidths and reasonable expectations concerning spatial resolution. Suppose, then, that one could build and maintain a DE, as a system capable of responding to any query about the surface and near-surface of Earth.

Now compare such a system to the abilities of one or more people able to observe and measure facts about the Earth's surface and near-surface through direct observation. Both

DE and the observers would be equally capable of responding to geographic queries. Moreover, it would be impossible to design an experiment to determine whether a response to a query came from a GIS and DE, or from an observer—that is, to determine whether the responding system was in contact with *bit* or *it* (Siegfried, 2000).

3.2 Naïve geography

The discussion thus far has been oriented to principles of scientific measurement, and has regarded the geographic world as a rigid, Newtonian frame. Science has always stressed the importance of shared understanding, and its role in communicating knowledge and in supporting the replicability of scientific results. But the human world does not necessarily respect such principles, and may instead be willing to accept the equal legitimacy of multiple, personal viewpoints. The research area known as Public Participation GIS (Craig et al., 2002) has investigated the ability of GIS to support multiple viewpoints; in the context of this chapter they might be regarded as multiple, equivalent versions of the same attribute, that is, $\langle x, z_1 \rangle$, $\langle x, z_2 \rangle$, $\langle x, z_3 \rangle$...

Egenhofer and Mark (1995) use the term *naïve geography* to describe a more human-oriented perspective, in which multiple points of view are possible, and in which the nature of geographic information reflects what people believe to be true, rather than what science has determined to be true. They define naïve geography as "the body of knowledge that people have about the surrounding geographic world."

Naïve geography creates interesting issues for the theoretical framework proposed here. On the one hand, multiple viewpoints can readily be accommodated through multiple tuples, as suggested above. But consider the statement "Santa Barbara is north of Los Angeles." The statement is widely believed to be true, and is reinforced by driving directions, since the main highway from Los Angeles to Santa Barbara is designated as US Highway 101 North. But if one takes eight compass directions, or even four, in neither case is the bearing of Santa Barbara in the North sector (between 337.5 and 22.5, or between 315 and 45, respectively). Instead Santa Barbara is better described as west of Los Angeles.

One might imagine building a GIS that respects such beliefs, but in reality it would be impossible to do so, because the geometric rules on which GIS is based, and which permit such queries to be addressed, form a mathematical system that is consistent with a small number of axioms. A GIS that respected such beliefs could not reason, because information could no longer be reduced to atomic form, and the rules on which reasoning is based would no longer be general.

4 TOWARDS A SEMANTIC THEORY OF GEOGRAPHIC INFORMATION

Consider the statement "Mount Everest is 8850m high." Symbolically, we might represent this as a tuple ⟨"Mount Everest", 8850⟩. But it fails to satisfy the earlier definition of an atomic tuple of geographic information because although 8850 is an instance of the attribute of elevation, "Mount Everest" does not directly identify a position on the Earth's surface. To satisfy the definition, it is necessary to know where Mount Everest is in some universal system of georeferencing, in other words to possess a second tuple

⟨x,"Mount Everest"⟩. Hence the tuple is of no value to someone who does not understand the term "Mount Everest," and is not geographic information until "Mount Everest"
is converted into a universal locator. A *gazetteer* is defined as a collection of tuples linking
placenames to universal locators.

The value of the tuple clearly depends on the receiver's knowledge of the location
of Mount Everest, but it also depends on what else is known about Mount Everest, that
location on the Earth's surface, and the elevation 8850. For example, one might know that
the location is one of rapid crustal movement and uplift; that an elevation of 8850m places
the summit in a zone dominated by strong jetstream winds; that summer precipitation in
the area is linked to the Monsoon, etc. All of these facts can be symbolized as tuples
linking either x or 8850 to other properties. Without them, the original tuple would be of
no value; and the more of them there are, the greater the value of the tuple, and the more
the tuple can support new reasoning, and successful response to new queries. The number
of tuples in which a concept appears might be used as a formal definition of *understanding*
of the concept.

It is now possible to sketch the broad outlines of the theory. First, information consists of linkages between properties or concepts; these atomic elements are known as
facts. Second, geographic information is composed of a particular type of fact in which
one of the constituent concepts is geographic location (and time if relevant). Geographic
location is specified in some universally understood method of spatiotemporal referencing, while the other concept in the pair is specified according to some convention that
is understood by both sender and receiver. Third, the value of a fact to a receiver depends on the number of other facts already possessed by the receiver, and in which one
of the constituent concepts is present; in other words, it depends on the receiver's level of
understanding of the concept. These basic principles are illustrated in Figure 2.

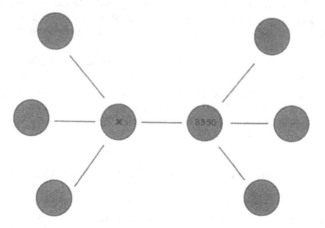

Figure 2: The value of a fact, represented by the connection between location x and elevation
8850, depends on the receiver's existing understanding of the two concepts, in the form of other
facts linking these concepts to additional concepts.

This simple framework is best illustrated with nominal concepts, but much more powerful reasoning is possible when concepts are ordinal, or interval/ratio. The elevation 8850, for example, allows naïve association with other facts about 8850, but it also allows operationalization of the concept of nearness, and reasoning about differences in height. Thus an elevation of 8840 is more similar than an elevation of 7850. Similarly since the elements of spatiotemporal location x are measured on interval scales it is possible to reason about nearby places, and to conduct advanced forms of spatial analysis that go far beyond the simple reasoning that is possible with nominal concepts.

5 CONCLUDING COMMENTS

The simple ideas outlined in this chapter seem to address the basic requirements outlined at the start: a theory of information that is independent of medium and structure, that gives formal meaning to such concepts as fact and understanding, that is invariant under reversible transformations, and that supports the determination of value. At its base is the notion that the communication of information and the construction of knowledge occur through a process of establishment of linkages between concepts.

The chapter has assumed that linkages either exist or do not exist, and has not dealt with the possibility of partial linkage. This is the basis of uncertainty, and the subject of theories of fuzziness and rough sets (Fisher, 2000, 2001), and expressed in English through phrases such as "might be" rather than "is". Statistics often formalizes such ideas through the concept of variance, and associated confidence limits. Consider, for example, the tuple $\langle x, 8850 \pm 2 \rangle$. One might argue that such a tuple is of less value than $\langle x, 8850 \rangle$, but of more value than, say, $\langle x, 8850 \pm 100 \rangle$. Moreover the theory of statistics allows inferences to be made from such statements, or from combinations of this and other uncertain statements.

How should one value such uncertain tuples? One possibility would be to assign fractional value based on the reduction of variance. For example, one could argue that without any other knowledge, the elevation of some point x could be anywhere in the full range of elevations exhibited by points on Earth. The uncertain tuple's value might then be expressed as $1 - s^2/S^2$ where S^2 denotes the observed variance of Earth surface elevations, and s^2 denotes the variance in the observation recorded in the tuple.

This chapter must be understood as work in progress: much more effort will be required to develop methods that can be used as a practical basis for valuing geographic information. Hopefully the ideas outlined here provide a useful starting point.

REFERENCES

Berry, B. J. L. (1964). Approaches to regional analysis: A synthesis. *Annals of the Association of American Geographers*, 54:2–11.

Cliff, A. D. and Ord, J. K. (1981). *Spatial Processes: Models and Applications*. London: Pion.

Craig, W. J., Harris, T. M., and Weiner, D. (Eds) (2002). *Community Participation and Geographic Information Systems*. New York: Taylor & Francis.

Csillag, F. and Kabos, S. (1996). Hierarchical decomposition of variance with applications in environmental mapping based on satellite images. *Mathematical Geology*, 28(4):385–405.

Egenhofer, M. J. and Mark, D. M. (1995). Naïve geography. In Frank, A. U. and Kuhn, W. (Eds), *Spatial Information Theory: A Theoretical Basis for GIS*, volume 988 of *Lecture Notes in Computer Science*, pp. 1–15. Berlin: Springer-Verlag.

Fisher, P. F. (2000). Sorites paradox and vague geographies. *Fuzzy Sets and Systems*, 113:7–18.

Fisher, P. F. (2001). Alternative set theories for uncertainty in spatial information. In Hunsaker, C. T., Goodchild, M. F., Friedl, M. A., and Case, E. J. (Eds), *Spatial Uncertainty in Ecology*, pp. 351–362. New York: Springer.

Goodchild, M. F. (2001). Metrics of scale in remote sensing and GIS. *International Journal of Applied Earth Observation and Geoinformation*, 3(2):114–120.

Goodchild, M. F., Egenhofer, M. J., Kemp, K. K., Mark, D. M., and Sheppard, E. S. (1999). Introduction to the Varenius project. *International Journal of Geographical Information Science*, 13(8):731–746.

Gore, A. (1992). *Earth in the Balance: Ecology and the Human Spirit*. Boston: Houghton Mifflin.

Güting, R. H. and Schneider, M. (1995). Realm-based spatial data types: The ROSE algebra. *The VLDB Journal*, 4(2):243–286.

Isaaks, E. H. and Srivastava, R. M. (1989). *Applied Geostatistics*. New York: Oxford University Press.

Kuhn, W. (1997). Approaching the issue of information loss in geographic data transfers. *Geographical Systems*, 4(3):261–276.

Longley, P. A., Goodchild, M. F., Maguire, D. J., and Rhind, D. W. (2001). *Geographic Information Systems and Science*. New York: Wiley.

Marchand, B. (1972). Information theory and geography. *Geographical Analysis*, 4(3):234–257.

Onsrud, H. J. and Calkins, H. (1993). NCGIA research initiative 4: The use and value of geographic information. Technical report, National Center for Geographic Information and Analysis, Santa Barbara, CA.

Peuquet, D. J. (1994). It's about time: A conceptual framework for the representation of temporal dynamics in geographic information systems. *Annals of the Association of American Geographers*, 84(3):441–461.

Shannon, C. (1949). *The Mathematical Theory of Communication*. Urbana: University of Illinois Press.

Siegfried, T. (2000). *The Bit and the Pendulum: From Quantum Theory to M Theory, the New Physics of Information*. New York: Wiley.

Sinton, D. (1978). The inherent structure of information as a constraint to analysis: Mapping thematic data as a case study. In Dutton, G. (Ed), *Harvard Papers on Geographic Information Systems*, volume 6. Reading, MA: Addison-Wesley.

Thomas, R. W. (1981). *Information Statistics in Geography*. Number 31 in Concepts and Techniques in Modern Geography. Norwich, UK: GeoBooks.

Tobler, W. R. (1970). A computer movie simulating urban growth in the Detroit region. *Economic Geography*, 46(2):234–240.

Worboys, M. F. (1995). *GIS: A Computing Perspective*. London: Taylor & Francis.

CHAPTER 3

Communicating Geographic Information in Context

Michael F. Worboys
National Center for Geographic Information and Analysis
University of Maine, ME, 04469, USA

1 INTRODUCTION

Underlying this chapter is the thesis that any foundational work on geographic information science, with the emphasis on information, should draw on general theories of information science. In particular, discussions of communication of geographic information need to be founded in the general work of information scientists on this topic. An example of a fundamental question that we would like to be able to begin to answer is "Given two data sources, is the information contained in one equal, greater, less than, or incomparable, with the other?". To approach questions of this sort, the notion of information 'value' becomes important. To discuss the question of the value of geographic information, Michael Goodchild (this volume) began with the commodity metaphor, where the value of geographic information is viewed in a similar way to the value of a barrel of crude oil. The issue then becomes how value is assessed. Simple-minded measures based upon quantities of data are clearly irrelevant, as the relationship between data and information is indirect. Goodchild goes on to argue for a measurement of value that takes account of the semantics of the information. This chapter takes the process one step further, beginning with the observation that the value of information also depends critically on the requirements and purposes of the recipient, which is part of the context in which the information is requested and received. To illustrate these points, consider the information conveyed by the statement "The window is unlocked". The value of this information will be greater for a burglar investigating the potential of a house in the neighborhood than for a casual passer-by, although the semantics in both cases could be the same. The key concept here is *relevance*, where relevant information is "information that modifies and improves an overall representation of the world" (Sperber and Wilson, 1995). The relevance of an item of information will depend not only on its syntactic form and semantic content, but also on the context in which it is transmitted and received.

The dominant theory of information flow has been the channel theory of Shannon (Shannon, 1948; Shannon and Weaver, 1949). This is discussed more fully in Section 2, but the basic structure is as shown in Figure 1. The concepts of transmitter and receiver of information are of course metaphors. The image of information leaving the source, being transmitted through a medium (channel) to a receiver, and on the way being the subject of degradation through noise, is so compelling that sometimes it is easy to forget that it is only an image, and an image with limitations. In particular, in Shannon and Weaver's elaboration of the metaphor, 'value' becomes synonymous with 'quantity' of information, which is measured by its capacity to 'surprise' the recipient. Surprise is measured using probability theory as the smallness of the chance that a particular signal is expected to be received. So, the smaller the chance, the greater the surprise, and the higher the information is valued. However, it is easy to find intuitive examples where value is not measured this way. Suppose our burglar has determined by observation that almost every night the owners of the house forget to arm the house alarm system, and only remember on those rare occasions when they takes their dog for a walk. On the night planned for the burglary, the lack of a dog being walked provides unsurprising but highly valuable information to the burglar. So, an understanding of context is key to the valuing of information; and more strongly, context is a key facilitator of information flow, without context there is no flow.

The classic work done by Shannon and Weaver establishes several very important concepts, in particular the notion of channel, which has close connections to our discussion of context. Later work by Dretske (1981), Barwise and Seligman (1997), Devlin (1991), and others, gives a formal approach to information content. This provides the motivation for the work below, where we follow these authors, and others, in extending the idea of an information channel in the sense of Shannon and Weaver to allow a notion of information in context. This is explored using examples from cartography, and focuses on properties of imperfection such as error, imprecision and vagueness. The purpose of this chapter is to provide a formal foundation for these ideas, with emphasis on modelling information source and destination as well as the channel that provides the context in which the information can flow. The chapter explores work on information flow, with particular reference to geographic information. We begin by explaining the principal constructs of classification, infomorphism and channel, and show how the idea of a representation can be approached using this apparatus. We draw examples from familiar representations of the geographic world, such as maps, but show how the framework developed here is quite general and may be extended to less familiar spatial representations (e.g. audio cues for in-car navigation systems). The chapter concludes by showing how contextual properties of representations, such as accuracy, precision and vagueness may be formulated, as well as suggesting some further avenues to explore.

2   SHANNON'S MODEL OF INFORMATION COMMUNICATION

Figure 1 shows the primary elements and relationships of the Shannon-Weaver theory of information communication. In this theory, a channel provides the dependency between source and destination, and this dependency is expressed as a probability that represents the 'surprisal value' (*entropy*) of the information transmitted. Entropy is about the amount of uncertainty that can on average be eliminated by a piece of information. Quantities that relate to the flow of information between source and destination, such as capacity, noise, and equivocation, are measured by probabilities, usually averaging over all possible eventualities. Thus the theory is quantitative, concerned with amounts of information, rather than information content. (See the chapters by Frank and Goodchild, in this volume, for more details on the formula for calculating information entropy).

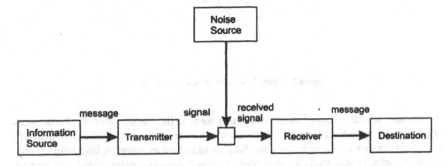

Figure 1: Shannon-Weaver model of information communication

Shannon-Weaver theory has been used as a basis for the calculation amounts of information conveyed by a cartographic data source. For example, Clarke (2000) defines the 'coordinate digital density function' as a metric based on entropy that allows feature-to-feature and map-to-map comparisons of spatial information quantity.

3 EXTENSION TO SHANNON-WEAVER THEORY

The rationale for the extended theory of information flow was provided by Dretske (1981) as the need to discuss the content of information communication, rather than just its quantity. Dretske also provides a critique of the concept of entropy as an appropriate measure of information quantity. His theory is summed up in the following four principles.

1. Information flow results from regularities in a distributed system.
2. Information flow crucially involves types and their particulars.
3. It is by virtue of regularities among connections that information about some components of a distributed system carry information about other components.
4. The regularities of a given distributed system are relative to its analysis in terms of information channels.

Principle 2 suggests classification of entities into types as the basis of a theory of information flow, and it is this notion that is now discussed.

3.1 Classification

A *classification* A is a triple $\langle A, \Sigma_A, \models_A \rangle$, where A is a set of objects of A to be classified, called *tokens* of A, Σ_A are the *types* of A used to classify the tokens, and $\models_A$ the typing relationship. A classification may be shown diagrammatically in Figure 2.

A relevant, simple example is a classification of part of the real world into a set of features. In this case the classification W consists of tokens that are real world objects and types that are features (e.g. churches, roads and mountains) that classify the objects.

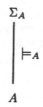

Figure 2: Classification A of tokens into types

Another classification of the same set of tokens is the classification L of real world objects based upon their location, either using some quantitative classification such as a coordinate system or a qualitative classification based for example on linguistic descriptions of spatial relationships to other objects. Some linguistic descriptions may be relatively clear-cut ('crisp'), for example, 'in England' (although even here there is sometimes room for doubt), while others, such as 'near the center of town' are less clear. Linguistic representations of classifications might well be vague, inaccurate, imprecise and context-dependent. One of the principal reasons for formalizing classification systems in this way is to understand such contextual properties.

Our third example demonstrates that a classification can represent spatial relationships as well as one-place spatial properties. Let W^* consist not only of tokens that are single real world objects, but also ordered pairs, triples, and so on. Then we can extend out classification to include binary and higher-order spatial relationships, so a binary spatial relationship such as 'Keele is in Staffordshire', can be formalized as (Keele, Staffordshire) $\models_{W^*}$ inregion, where inregion is the type indicating the spatial relationship. A system such as the region connection calculus (RCC) (Randell et al., 1992), provides a simple formal classification of binary spatial relationships between regions. This example shows that a classification scheme can have its own internal structure. Another example of extra structure in the classification scheme is provided by a hierarchical feature classification, such as may be found in the CORINE data set, which maps land cover units with a minimum spatial granularity of 25ha based on 44 land cover classes arranged in a 3-tier taxonomy (Bossard et al., 2000).

A classification scheme has its internal logic, arising from the classification of tokens into types. A set of inferences may be associated with a classification, expressed as $X \rightarrow Y$, where X and Y are sets of types. For the feature classification example, {church, library} ⊢ {building} (or in abbreviated form church, library ⊢ building) expresses the fact, noted extensionally from empirical observation of churches and libraries, or intensionally from the class definitions, that every church or library is a building (see Section 3.4). A key idea is that information flow is the movement of inferences from one classification to another.

3.2 Infomorphism

A classification provides a model of an information system, whether human or machine, source, destination or mediator. The next part of the theory provides a mechanism for relating information systems having similar informational structure. Figure 3 shows the formal way, known as an *infomorphism*, of structurally relating two classifications. Notice that the functions f^+ from types to types and f^- from tokens to tokens work together contravariantly. In order that classification structure is preserved, the pair of functions f^+ and f^- satisfy the following condition.

$$\text{For each token } b \in B \text{ and type } \alpha \in \Sigma_B. \ f^-(b) \models_A \alpha \text{ iff } b \models_B f^+(\alpha) \qquad (1)$$

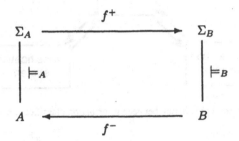

Figure 3: Infomorphism f from **A** to **B**

A good example of an infomorphism is provided by the relationship between a map and the part of the real world that it represents. In this case, let A be a set of map elements and Σ_A be the set of map symbol types to which the map elements may belong. Let B be a set of real world entities and Σ_B be the set of feature types that the real world entities may belong. Condition 1 gives the infomorphism constraint that ensures that the map is set up correctly in relation to the world. For example, if b is an entity in the real world and $f^-(b)$ is a particular red line on a map, then condition 1 gives $f^-(b) \models_A$ redline iff $b \models_B$ road. So, this infomorphism will be satisfied if map elements corresponding to roads are shown as red lines.

Of course, in reality maps and the process of cartography are much more complex than this example indicates. For example, the correspondence between real world and cartographic artifact may be imperfect in several different ways. To model this more complex scenario, and to explore the role of context (in this case the cartographic context of map production), we need the concept of 'channel', to be developed next.

3.3 Channel

As in Figure 4, the concept of channel provides a mediator between several information classifications, and will be used to model the context in which transmission of information takes place. Figure 5 shows the general configuration in the binary case. The *channel core* **C** acts as a medium for the transmission of information between **A** and **B**. The tokens of **C** are called *connections*, where a connection $c \in C$ *connects* $f^-(c) \in A$ with $g^-(c)$. Continuing our earlier example, if **A** consisted of map elements and their types (e.g., thick red line) and **B** consisted of real world transportation elements and their types (e.g. road), then the connection might be made by c, the cartographic act of representing a particular

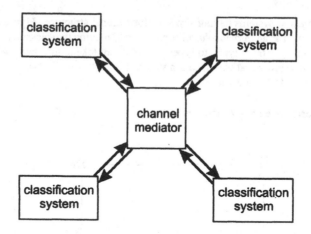

Figure 4: Mediation between information classifications

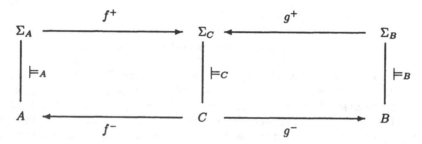

Figure 5: Channel

road as a thick red line on the map. Note that a channel makes connections both at the token and type levels, by functions f^- and f^+, respectively.

Channels allow information flow, and in the general theory may be n-ary. However, only binary channels are needed for this discussion. We will return to channels in Section 4, when we will see how they are used to make precise the notion of a representation, using as an example cartographic representations, and in Section 6 where we see channels used to model multi-contextual situations.

3.4 Local logics

Each classification has a logic associated with its type system, called its *local logic*. As we will see later, it is the local logic that holds the informational structure of the classification system, and movements between local logics indicate the information flow. Local logics allow inferences in the type system, and will be represented using sequents, in the style of Gentzen (1969). For types α, α', the sequent $\alpha \vdash \alpha'$ indicates that the inference from α to α' is valid. For example, suppose that the types were real world features, then the sequent **house ⊢ building** indicates that houses are buildings. The sequents in the local logic are called *constraints*.

Consider the classification $A = \langle A, \Sigma_A, \models_A \rangle$. The standard case is that, for types $\alpha, \alpha' \in \Sigma_A$, the sequent $\alpha \vdash \alpha'$ is valid if and only if

$$\forall a, a' \in A. \ a \models_A \alpha \text{ implies } a' \models_A \alpha'$$

So, for our example, **house ⊢ building** would be valid if all real world house objects in the system are buildings. A token is called *normal* if it satisfies all constraints in the logic.

In general, not all tokens may be normal. In our example, suppose that there exists a real world object that is a house but not a building. These non-normal exceptions may be used to model kinds of imperfection in the informational structure and flows, such as inaccuracy and incompleteness. They are considered in Section 5

4 REPRESENTATION

The reader who has been patient enough to follow us to this point will we hope now begin to see some rewards for the pain of all this formal machinery, in that it allows us to capture with considerable richness the notion of *representation*. The notion of representation is central to geographic information science, with the map as its paradigm.

Formally, a *representation system* consists of a binary channel (see Section 3.3), on the core of which is imposed a local logic. Consider the binary channel **C** shown in Figure 5. Suppose that the channel core has a local logic associated with its types. Call the classification **A** the *source* and **B** the *target* of the representation system.

The *representations* are the tokens of **A**. We say that *a is a representation of b* if there is a connection from a to b. A type α in the source *indicates* type β in the target if $f^+(\alpha) \vdash g^+(\beta)$ in the local logic of the core.

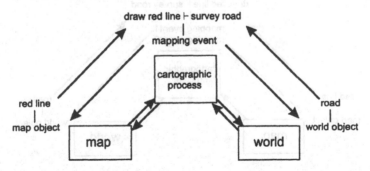

Figure 6: The cartographic process as mediator between map and world

As an example, let **A** be the classification of map elements and **B** be a set of real world objects and features. Then, as in Figure 6, the representations are the map objects, and map objects are representing real world objects. Let the core **C** be the cartographic process of surveying real world objects and associating them with map elements. The cartographic process might contain rules such as associating a red line on the map with a road on the ground. So a constraint in the local logic might be draw red line ⊢ surveyed road. In that case, using our definition of 'indicates', *red lines indicate roads*. In terms of information flow, 'this map object is a red line' carries the information that 'the corresponding real world entity is a road', relative to correct survey and cartographic conventions. In summary, the channel of cartographic process provides the context in which the map captures and represents information about the world.

The treatment here is necessarily brief, but shows the possibilities of formally modelling flows between geographic representations, e.g. between graphical and textual representations of geographic phenomena. The theory of representations is developed in a general setting by Barwise and Seligman (1997).

5 PROPERTIES OF REPRESENTATIONS

5.1 Accuracy

A representation is *accurate* if there is an appropriate correspondence between the representation and the target domain being represented. The theory of representations, the beginnings of which were developed in the preceding section, gives us a succinct way of characterizing accuracy. We will use the representation system provided by the channel in Figure 5, with a local logic imposed upon its core.

We have the following definition: The token $a \in A$ is *an accurate representation of token* $b \in B$ if and only if

1. a is a representation of b.
2. The token that connects a and b is normal.

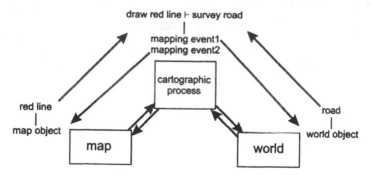

Figure 7: An error matching tokens in a map representation of the world

Returning to our 'maps of the real world' example, with the cartographic local logic containing cartographic conventions, including the constraint draw red line ⊢ surveyed road. Figure 6 shows an accurate representation in the map of the world. The token c that connects this representation to the real world object is normal, as it satisfies all constraints in the local logic, in particular the constraint draw red line ⊢ surveyed road.

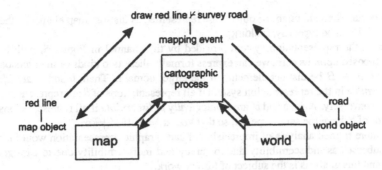

Figure 8: An typological error in a map representation of the world

The next two figures illustrate how two types of error can occur. Figure 7 shows a violation of condition 1 above. In this case, no connection is made, through a proper cartographic process, between a particular road and a particular red line on the map (maybe a different red line on the map is used to represent the road) . So the road is inaccurately represented in the map. On the other hand, Figure 8 shows a different form of error, resulting from a violation of condition 2 above. In this case the token c that connects this representation to the real world object is not normal, as it does not satisfy the constraint that draw red line ⊢ surveyed road. An error of typology will result (maybe the road will be represented as a railway). While this example is simple, it points the way to analysis of the accuracy of complex representations, including verbal forms, in which the local logics are more interesting.

5.2 Precision

Precision refers to the level of detail that a representation captures from the target domain, and is independent of accuracy. Objects that are discernibly distinct in the target domain may or may not be differentiated in the representation domain, resulting in various degrees of precision. Distinction between entities may occur in two ways: their tokens may be distinguished (in an information system this could be the result of distinction between object identities), or their types may be distinguished (class membership would serve to do this in an object oriented information system). Put more succinctly, we have two forms of indiscernibility that may result from an imprecise representation:

identity indiscernibility The representation fuses the identities of objects in the target domain.
typological indiscernibility The representation fuses the types of objects in the target domain.

In terms of the theory developed here, differentiation of tokens occurs by means of their types, so imprecision will result if two tokens with distinct type sets in the target domain have the same type sets in the representation domain. In our mapping example, a case of identity indiscernibility occurs when a collection of distinct buildings (two churches and a school, say) in the world are merged into a single map object. Typological indiscernibility might occur when although the objects are represented as distinct map elements, they are assigned the same type, say, building.

Using the representation system provided by the channel in Figure 5, with a local logic imposed upon its core, we can express formally these two kinds of imprecision. Let tokens $b, b' \in B$ be distinct elements of the target domain. Then, b and b' are *identity indiscernible* in the representation system if all representations of b are representations of b', and conversely. Also, b and b' are *typologically indiscernible* if all representations a of b and a' of b' have the same type set, so that $\forall \alpha. \ a \models \alpha$ iff $a' \models \alpha$.

A more refined analysis of imprecision of cartographic representation would include such subcases as indiscernibility due to survey and indiscernibility due to cartographic rules, and this analysis is the subject of further work.

5.3 Completeness

A desirable property of a representation is that it represents all the content of interest in the target domain. As with the other properties, there are forms of completeness relating to types and tokens. Using our example to illustrate one case, the representation in Figure 9 is clearly incomplete at the token level, as there are entities in the real world which are not represented by map entities.

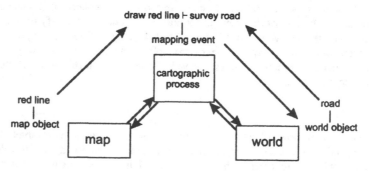

Figure 9: An incomplete map representation of the world

6   VAGUENESS AND CONTEXTUALITY

Vague predicates admit borderline cases for which it is not clear whether the predicate is true of false. Vague entities have parts for which it is unclear whether or not the part is part of the entity (Williamson, 1994; Varzi, 2001). There are many examples of vagueness in geographic phenomena. An example of a vague predicate is 'nearness', and examples of vague entities are a mountain or 'the north of Maine'. The paradox of the heap demonstrates that vagueness is not easy to handle using classical reasoning approaches, and many more or less successful attempts have been made to provide a principled account of the semantics of vagueness, and reasoning with vague notions. Vague notions often have associations with context dependence and subjectivity. What may count as the north of Maine for me may be different for you (*subjectivity*). What may count as near home may depend on whether I am thinking of walking of taking the space shuttle *context dependence*.

Even though classical approaches to reasoning have difficulties with vagueness, that does not mean that it is impossible to have information flow with vague ideas. For example, if you phone me to tell me that you are currently in the north of Maine, and I know that I am a few miles to the north of you and in Maine, then I can reply to you that I am also in the north of Maine, *even if our notions of the north of Maine are quite different.*

We will provide a sketch of how vagueness may be discussed using the theoretical ideas so far developed, and briefly show how the theory can be used to model information exchange in the cases where the context involves vague notions. We will use as a running example the vague spatial relation of nearness (Worboys, 2001; Fisher and Orf, 1991). The point that we are emphasizing here is that the theory of information flow allows us to reason with context dependent ideas, of which nearness is a good example.

Suppose we have two agents, A and B, with their own (and possibly quite different) notions of nearness. We will set up classification systems connected by a channel, as in Figure 5, to model the nearness notions of A and B, whose contexts are mediated by the channel C.

The classification systems A and B, modelling the nearness notions of agents A and B, are set up as follows. Let the tokens of A be ordered pairs of locations. Let Σ_A consist of the type: near. The idea of the A classification system is that:

$$\forall a = (l, m) \in A. \ a \models \text{near iff agent A believes } l \text{ is near } m \qquad (2)$$

We have a similar set up for the classification system B. Let the tokens of B be ordered pairs of locations and Σ_B consist of the type near. Then:

$$\forall b = (l, m) \in B. \ b \models \text{near iff agent B believes } l \text{ is near } m \qquad (3)$$

Of course, because of the subjectivity and context dependence of nearness, these two classification systems will in general be different. What they have in common is expressed by the mediating channel C. As an example of a method for constructing channel C, we need some preliminary definitions.

Definition
For points $x, y \in \Re^2$.
line$(x, y) = \{z \in \Re^2 \mid z \text{ is on the closed straight line segment } xy\}$

Definition
For a point $x \in \Re^2$, a *neighborhood of* x is a region S of $\Re^2$ with the following properties:

1. $x \in S$
2. If $y \in S$, then line$(x, y) \subseteq S$

The idea is that neighborhoods in $\Re^2$ model the idea of a region of locations that are near to a particular location. The second property of neighborhoods is an attempt to model the idea of 'downward closure' with respect to nearness. We can note that every neighborhood is connected in $\Re^2$, but not every connected region is a neighborhood.

We are now ready to define the mediating classification system **C**. Let the tokens of **C** be points $x \in \Re^2$. Let Σ_C consist all the neighborhoods of the origin $\Re^2$.

If we assume that agent **A** has a 'consistent' notion of nearness (e.g. we might assume that if agent **A** believes x is near, and y is nearer than x then **A** should also believe y is near), then it will be possible to set up an infomorphism from **A** to **C**. Similarly, we should be able to set the infomorphism from **B** to **C** (Note that in general it is *not* possible to set up an infomorphism directly from **A** to **B**). We now have a configuration in which limited information can flow between agent **A** and agent **B**, using the channel **C** that expresses some common geometrical understanding about nearness.

7 CONCLUSIONS AND FUTURE DIRECTIONS

This chapter has discussed the application of some of the recent work on information flow to geographic information. It has argued for an approach that would handle information content and context rather than quantity. The theory of information flow, due to Dretske, Barwise, Seligman and others, has been developed with reference to examples from geographic information. We have shown how it can be used to examine a range of issues involved in representation of spatial phenomena, including accuracy, precision and incompleteness. We have also used the theory to discuss the key issue of context, and shown how multi-contextual situations can be approached, using the example of the vague spatial relation 'near'.

The work has proceeded by examining ways in which the transmission through a conduit metaphor of information communication can be extended beyond that of the foundational work of Shannon and Weaver. However, there has been some criticism of this metaphor as unnecessarily limiting, and not accounting for content, context, time, and the nature of the transmission medium. Some of these concerns have been addressed by authors previously cited, and the present work, but others remain. A more direct image is provided by Sperber and Wilson (1995), where communication of information is defined as a process where "one device modifies the physical environment of the other. As a result, the second device constructs representations similar to representations already stored by the first device".

As we have seen the information content of an information source depends on the channel and destination (that is, the delivery context), as well as the source context. So, one of the key issues that arises is the need for further work on context, how it is specified, and how it affects information flow. While this chapter indicates some beginnings, it seems clear to us that the theory developed here can be used to address several key questions in geographic information science, including the following.

- How can we determine the informational content of a spatial dataset, taking account of source and delivery contexts?
- Can we determine whether two spatial datasets contain the same information?
- How can we compare the informational content of two different representations (say, graphic and linguistic) of geographic phenomena?
- How can we represent and reason with information arising from vague representations?

One of the neat aspects of the work presented above is the importance it places on the role of mediator in the flow of geographic information. This mediator can, for example, be a map, in the case of flow between two humans, or the cartographic process, in the case of the representation of real world phenomena in a spatial data set. In this way, the approach allows detailed consideration of the role of the mediation process, as context, in information delivery.

REFERENCES

Barwise, J. and Seligman, J. (1997). *Information Flow*. Cambridge, England: Cambridge University Press.

Bossard, M., Feranec, J., and Otahel, J. (2000). CORINE land cover technical guide. Technical Report 40, European Environment Agency, Copenhagen.

Clarke, K. (2000). The coordinate digit density function as a tool for analytical cartography. In *Proceedings of GIScience2000, Savannah, GA*.

Devlin, K. (1991). *Logic and Information*. Cambridge, England: Cambridge University Press.

Dretske, F. (1981). *Knowledge and the Flow of Information*. Cambridge, MA: Bradford Books, MIT Press.

Fisher, P. and Orf, T. (1991). An investigation of the meaning of near and close on a university campus. *Computers, Environment and Urban Systems*, 15:23–35.

Frank, A. U. (2002). Pragmatic information content—how to measure the information in a route description. In Duckham, M., Goodchild, M. F., and Worboys, M. F. (Eds), *Foundations in Geographic Information Science*, pp. 47–68. London: Taylor & Francis.

Gentzen, G. (1969). Investigations into logical deduction. In *The Collected Papers of Gerhard Gentzen*. Noth-Holland.

Goodchild, M. F. (2002). The nature and value of geographic information. In Duckham, M., Goodchild, M. F., and Worboys, M. F. (Eds), *Foundations in Geographic Information Science*, pp. 19–31. London: Taylor & Francis.

Randell, D. A., Cui, Z., and Cohn, A. (1992). A spatial logic based on regions and connection. In Nebel, B., Rich, C., and Swartout, W. (Eds), *KR '92. Principles of Knowledge Representation and Reasoning: Proceedings of the Third International Conference*, pp. 165–176, San Mateo, CA. Morgan Kaufmann.

Shannon, C. (1948). A mathematical theory of communication. *The Bell System Technical Journal*, 27:379–423, 623–656.

Shannon, C. and Weaver, W. (1949). *The Mathematical Theory of Communication*. Urbana: University of Illinois Press.

Sperber, D. and Wilson, D. (1995). *Relevance: Communication and cognition*. Oxford: Blackwell, 2nd edition.

Varzi, A. (2001). Vagueness in geography. *Philosophy & Geography*. Forthcoming.

Williamson, T. (1994). *Vagueness*. London: Routledge.

Worboys, M. (2001). Nearness relations in environmental space. *International Journal of Geographical Information Science*, 15(7):633–651.

CHAPTER 4

Pragmatic Information Content—How to Measure the Information in a Route Description

Andrew U. Frank

Institute for Geoinformation, Technical University of Vienna
Vienna, A-1040, Austria

1 INTRODUCTION

Shannon and Weaver published in 1949 a breakthrough book on how to measure the information transferred over a channel. They introduced the unit *bit* as a measurement unit for information, which stands for one binary decision. This method is commonplace today and widely used to measure amounts of *data* capacity for storage devices, etc. It does, however, not assess the *pragmatic information* content of a message.

Two messages of very different data and size may communicate the same message and have therefore the same information content; we will call this the pragmatic semantics. We also know that the same message may have very different information content for different users. A theory for a measure of pragmatic information content must account for the fact that different messages may have the same content and that the same message may have different content for different recipients.

In the prototypical situation a recipient of a message uses the information to make a decision about an action. Other situations, where information is assimilated for later usage require some slight extension of the method, but always, information is only useful pragmatically when it influences a decision.

To determine pragmatic information content, the user is modelled as an algebra. All messages which lead to the same actions have the same information content, which is the minimum to determine the action. If two users differ in the action they consider, their algebras differ and therefore the information they deduce from the information content of the same message is different. Both cases are formalized in this paper with algebraic tools.

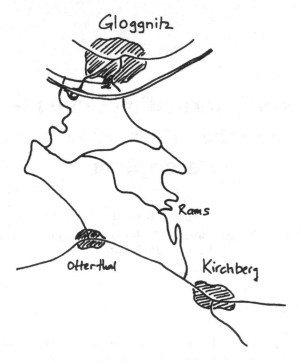

Figure 1: Map of the area

1.1   Motivation example

A friend tells me how drive from *Kirchberg am Wechsel* to *Gloggnitz* (Figure 1)—a drive between two small towns south of Vienna (Table 1):

Kirchberg am Wechsel to Gloggnitz
Follow the road to Otterthal
In Otterthal turn right towards Gloggnitz
Follow the road through Schlagl and Graben
Cross under the Semmering highway
Follow the road into the town of Gloggnitz

Table 1: A friend's driving directions from Kirchberg am Wechsel to Gloggnitz

I do not fully trust his information and check with a service on the Web, which produces the following route description (Table 2):

Your route from Kirchberg am Wechsel to Gloggnitz:
The total distance is 13.1 km.
To drive this distance will probably take 00:21 (hh:mm).

Street name	Driving Time	Route Description	Length	Distance from start
LH134\ Markt	00:00	On LH134\ Markt	4,1 km	4,1 km
LH134\ Otterthal	00:06	Turn right on LH134\ Otterthal	6,6 km	10,6 km
LH134\ Graben	00:16	Turn right on LH134\ Graben	430 m	11,0 km
LH134\ Graben	00:16	Turn right on LH134\ Graben	770 m	11,8 km
B27\ Semmeringstrasse	00:18	Turn right on B27\ Semmeringstrasse	650 m	12,5 km
Hoffeldstrasse	00:19	Turn right on Hoffeldstrasse	500m	13,0 km
Sparkassenplatz	00:20	Turn right on Sparkassenplatz	50 m	13,0 km
Sparkassenplatz	00:21	Turn left on Sparkassenplatz	128 m	13,1 km

Table 2: Web driving directions from Kirchberg am Wechsel to Gloggnitz

Is this the same route as described by my friend? My curiosity is started and I check two other descriptions (Tables 3 and 4).

I realize that I have received four times information to drive between the same locations—encoded in four different forms. Is it the same information? What do we mean by "the same information"? Careful analysis shows that the first two descriptions (Tables 1 and 2) give the same route and differ from the last two (Table 3 and 4). The instructions contain the same information but present it in a different form. How do we measure *pragmatic information content* for messages of different size, which lead to the same actions?

1.2 Analysis

The theory of Shannon and Weaver (1949) is widely used to measure the size of messages in storage or transfer; it measures the amount of data which is stored or transmitted in *bits*, i.e., a unit of a single binary decision. It does not measure the pragmatic information content of a message—it measures the amount of data in a message, not the effects the message has.

The pragmatic information content depends on the message and the situation in which the information is used to make a decision. It is therefore necessary to model the receiver of the message and the decision the message is used for making. The message itself is assumed to be a fixed artefact produced by the sender. As such its content after production does not depend on the sender anymore. Practically, the interpretation of a message by

Start: A-2880 Kirchberg am Wechsel
Destination: A-2640 Gloggnitz

No.	State	Node	Direction	Road	km	Total km	Time
1	A	Kirchberg am Wechsel			0.0	0.0	00:00
2	A	Ramssattel			2.7	2.7	00:15
3	A	RS	Left on	B27	6.5	6.5	00:16
4		Gloggnitz			1.5	10.7	00:18

Total distance: 10,7 (km); total driving time: 00:18 (hh:mm)

Table 3: Alternative driving directions

Time	Total km	Description	Turn	Road
00:00	0.0	A-2880 Kirchberg am Wechsel – Markt		
00:13	6.2		Half left	
00:16	7.8		Stay left	
00:22	11.7		Turn right on	B17
00:22	12.0	A-2640 Gloggnitz		

Table 4: Alternative driving directions

a receiver may be affected by the receiver's knowledge about the circumstances of the sender.

In this article, I suggest a formal approach to relate data to the practical situation in which it becomes information. When a message is used to decide on some action, then the message becomes information (in the sense of pragmatic semantics) and the pragmatic information content of a message can be identified and measured—*with respect to this decision context* (Table 5). The information content of all equivalent messages (rule EQ) is measured as size of the minimal message necessary for the decision (rule SAME); the pragmatic information content is measured against a practical situation in which the information is used—the same message has different pragmatic information content for different users and different uses (rule DIFF). The use of the information is formalized as an algebra and the size of the minimal message is measured with the method of Shannon and Weaver.

This article follows from ontological studies with a multi-tier ontology (Frank, 2002, 2001a,c); and expands on an idea by Wittgenstein where he suggests to use games as an analogue, which abstract important properties from practical situations which are too complex to analyze. In my contribution to the Wittgenstein-Symposium 2001 (Frank, 2001b) I have explored the formalization of board games as algebras and these concepts are here extended to the formalization of a user of information in a spatial decision: driving a car on a road network is comparable to a board game. This leads me to the definition of a pragmatic information measure, which fulfils the two conditions mentioned above.

Theory of pragmatic information content	
(EQ)	Two messages are equivalent when they lead to the same actions.
(SAME)	Equivalent messages of different size have the same pragmatic information content.
(DIFF)	The same message has different pragmatic information content when used in different decision contexts.

Table 5: Pragmatic information content of messages

A connection of this theory to game theory von Neumann and Morgenstern (1944) is possible.

The paper is motivated by the need to measure the information provided by Geographic Information Services, like the route planners initially shown (Krek, 2002). How should such services charge for the information they provide? By the character transmitted? By connect time?

Unfortunately, actual route descriptions as given in the examples leave many questions for a driver open. They are difficult to use and it is not clear what their intended semantics are. In this article a formal description of semantics of standardized route descriptions are given and types of route descriptions with their semantics defined. For simplicity, I use as background instructions for navigating in a city street network already published elsewhere (Frank, 2000). To measure the pragmatic information content of other messages follows the same concepts, but results in more variability in the content, introduced by more variability in the decision the information could be used for. Different situations lead to different information extracted from the same message.

The paper is structured as follows: the next section reviews the classical theory of information measurement and the following Section 3 describes pragmatic information content measure informally. Section 4 shows how to model the decision context of a user using a message to make a decision as an algebra. Section 5 models different user situations as algebra. Section 6 then defines a measure which satisfies the equations listed. Section 7 points to application of these ideas to the geographic information business. The concluding Section 8 summarizes the results and points to some open questions.

2 THE MATHEMATICAL THEORY OF COMMUNICATION

In their landmark contribution Shannon and Weaver have analyzed the transmission of messages over channels and how the message is affected by noise. Their measure of information is applicable to the technical level of communication. It measures the size of a message in binary decisions necessary to reconstruct the message and suggested *bit* as the fundamental unit to measure information content. This measure is widely used today and the unit bits and its multiples, i.e. Byte = 8 bits, and kilobytes, megabytes, etc. have become household words to measure the capacity of storage devices and communication channels.

A message of one bit is transmitted over a channel from a sender to a receiver if the sender informs the receiver about a decision between exactly two choices of equal probability; in the prototypical case the sender throws a coin and transmits the result as 'heads' or 'tails'. To decide between more choices—e.g., the selection of a candidate in an election out of eight requires three binary decisions (first to select the first or the second four, then the first or second two out of the four and then one out of the two). In general, the information content in bits is the logarithm to base 2 (logarithms dualis, *ld*) of the number of choices. For practical purposes the result is usually increased to the next entire number.

$$(entropy) \ H = -K \sum_i p_i * ld \ p_i \tag{1}$$

If the choices are not of equal probability, then the information H is the weighted sum of these probabilities (entropy formula). The negative sign is necessary to convert to a positive value; notice that the probabilities p_i are all less than 1 and that $ld \ p_i$ therefore negative. K is a positive constant. Shannon pointed out the relationship with similar measures in physics and suggested the term *entropy* (or uncertainty) for this property of a source of messages.

To guard against errors in transmission over noisy lines, *redundancy* is added. Redundancy can be used to reconstruct a partially transmitted text and to correct transmission errors. Typically natural language text contains considerable redundancy, estimated for English at about 50%. A text where every other character is left out can be read without much trouble.

A message can be encoded with different redundancy—usually the redundancy will be matched such that the signal and the redundancy are less than the capacity of the channel. Redundancy is measured in bits as well. The size of a message as transmitted is therefore the data content plus redundancy. Given only a message, one can measure the size of the message in bits, but not separate the data content from the redundancy. The next section discusses a method to identify pragmatic message content and separate it from redundancy.

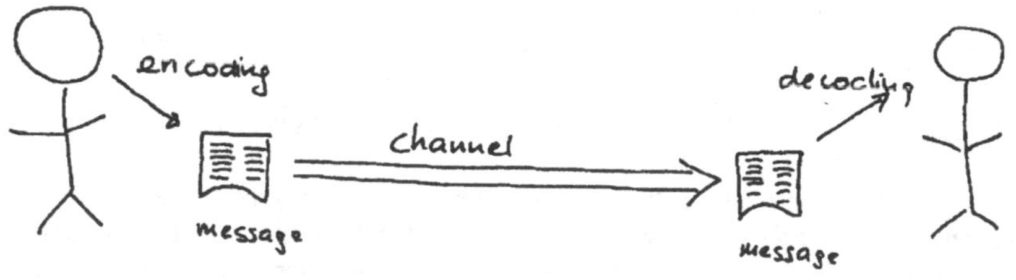

Figure 2: Transmission of a message through a channel (Shannon and Weaver, 1949)

3   PRAGMATIC INFORMATION CONTENT

Pragmatic semantics and pragmatic information content of messages must be investigated not in a transmission situation as described by Shannon and Weaver (see Figure 2) but in a decision situation (Figure 3). The connection between the information in the message which is used to make a decision about some action and the decision itself needs to be considered—Shannon and Weaver's method stops when the message is correctly received.

Figure 3: The decision context

Information is used to make decisions between actions—it is difficult to see another use of information. Often, we acquire information ahead of time and store, i.e. learn, facts for which we expect later a use in an expected decision situation. The determination that the four messages in the initial example are essentially the same information is based on the pragmatics of 'finding my way to my friend's town'. The messages are equivalent if I find the same way to my friend's home.

A measure of pragmatic information content is different from the measure of data size of messages using the theory of Shannon and Weaver. The measure of Shannon and Weaver is not adequate for information content. For example, two of the route descriptions given initially have the same pragmatic semantics, but different message sizes. A widely held opinion therefore wants to restrict the entropy formula to technical circumstances and declares it inappropriate as a 'real' information measure.

3.1   Pragmatic equivalence of messages

Messages have the same pragmatic semantics if they lead to the same action—assuming a fixed decision situation. If I have to drive from Kirchberg to Gloggnitz, then a series of decision situations are fixed: at each intersection I have to decide which way to turn. Two of the instructions given initially, if properly interpreted, lead at these intersections to the same decisions. These instructions are therefore pragmatically equivalent.

3.2   Different messages for different decision contexts

If we give instructions, we adapt them to the person to whom we give them. Route
descriptions assume that drivers have certain abilities. Some route descriptions refer to
cardinal directions, most web-based ones use distances. Not all drivers are certain where
the cardinal points are while driving and many ignore the odometer which would give
them distance information. They cannot effectively use such instructions. Some drivers
can follow a named or numbered highway through many intersections; others need instructions at each intersection. I once went in Virginia from Lee Highway 2000 to Lee
Highway 10620—14 miles of winding road through many tricky intersections where I got
lost more than once! Many route descriptions from the Web require additional information gathered from the road signs and knowledge about the location of places mentioned
on road signs and in the route description.

An instruction type is geared towards a specific decision situation, where the decision
maker has determined ability and knowledge. Users with more knowledge can often use
instructions prepared for less knowledgeable users, but not the reverse. For example, users
with a general geographic knowledge of the area can use detailed descriptions, ignoring a
large part of the message.

3.3   Pragmatic information content

If two messages are pragmatically equivalent—i.e., they lead to the same decision—they
have the same pragmatic information content. Even if their size, measured as size of
data to be transmitted using the entropy formula, is different, the measure of pragmatic
information content must be the same.

For a knowledgeable user (agent C in Figure 4) a succinct instruction is sufficient with
a low pragmatic information content. If the same user is given a more detailed one, for
him, the more detailed instructions have the same pragmatic semantics and therefore the
same (low) pragmatic information content. For a user who requires a detailed instruction
(agent B in Figure 4) the same message has a higher pragmatic information content because he has less world knowledge already available. Other agents acquire information
from the environment and need therefore fewer instructions (agent A in Figure 4). For
the knowledgeable user, much of the detailed message is redundant and not part of pragmatic information content—in the extreme case, where somebody knows the way from
Kirchberg to Gloggnitz already, the message does not contain any new information, i.e.,
no pragmatic information content. The decisions taken without the instructions would be
exactly the same!

3.4   Formal description of use of information in decision required

Pragmatic information content can only be measured with respect to a determined decision situation and decision process. It is therefore crucial to define the decision context
precisely and to assess instructions with respect to the decision context. This will be done
formally in the next two sections.

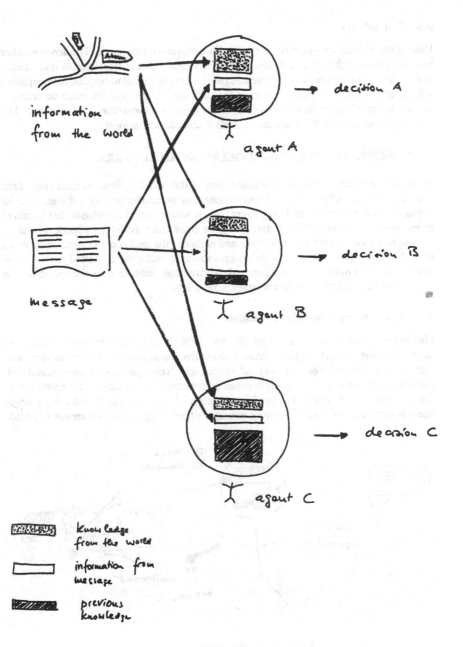

information from the world

message

knowledge from the world

information from message

previous knowledge

decision A

agent A

decision B

agent B

decision C

agent C

Figure 4: Three different agents use different amounts from information resources

3.5 Redundancy

Data in the instruction which is not required is considered redundant. The driver reaches his target without this data as well—only the necessary part is translated to information and used to make the decision. In practice, redundancy is crucial to respond to unexpected situations, missing street signs, errors in the data used to produce the route description, etc. In this article only the role played by the necessary information is investigated. The value of redundancy in the instructions needs a separate assessment.

4 A DECISION CONTEXT IS MODELED AS AN ALGEBRA

To determine information content a description of the decision situation must come first. This description explains how the instructions can be understood by a driver, i.e., the semantics of the instructions. Using agent theory, which considers autonomous agents in an environment (Ferber, 1998; Weiss, 1999), we construct a model of an agent simulating driving in a model of the street network and consider the decision this agent must make at each intersection. The agent with the operations to make the simulated moves in the street network is modelled as an algebra; the instructions must identify the operations the agent must take and provide the necessary parameters.

4.1 Agent theory to model the situation

Multi-agent theory gives a framework (Ferber, 1998; Weiss, 1999) in which we can formalize decision contexts: Agents perceive an environment, make decisions based on their perception and knowledge, and carry out actions which change the environment and their position in it (Figure 5). This cycle is executed repeatedly, for example, for each instruction in a route description. For the formalization of simple decision contexts, a single agent is sufficient and other interesting concepts of multi-agent systems are not required.

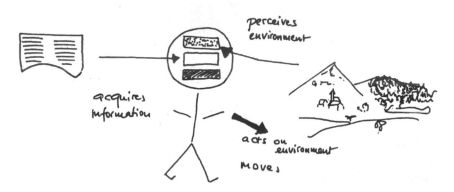

Figure 5: Cognitive agent in environment

Multi-agent programming extensive tools are available (Rao, 1996); our interest here is in building a method for measuring pragmatic information content and we revert to mathematical tools, in this case algebra. Multi-sorted algebras consist of types (mathematicians call them sorts), operations, and axioms. The operations have objects of the defined types as arguments and produce such objects; the axioms describe the outcome of the operations (Birkhoff and Lipson, 1970; Loeckx et al., 1996).

Algebras have the desirable property that they are semantically self-contained and do not require other definitions; an algebra defines objects and operations completely and independently (up to an isomorphism).

4.2 Ontology: The street network assumed

It is not evident how to interpret the route descriptions shown initially and it is difficult to see if the descriptions are equivalent. One would have to follow them actually driving and then see if the path taken is the same. For the formal treatment here, I use a simplified description of a small part of downtown Santa Barbara (Figure 6); this has been used previously as a simplistic environment for map making and map interpreting agents (Frank, 2000).

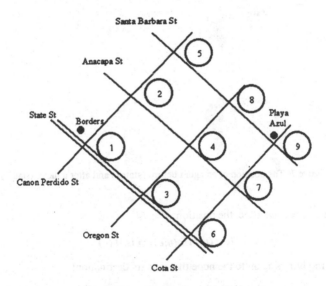

Figure 6: A small subset of streets of downtown Santa Barbara (with intersection identifiers)

The street network consists of street segments, which run from an intersection to the next. In the model the street intersections are identified with numbers. This is a simplification of the well-known TOURS model (Kuipers and Levitt, 1978, 1990). The state of the world and the agent is merged in a single state variable, which maintains the complete state of the model of agent and environment.

```
class BasicDrivingAgent agent env intersection where

    startAt :: intersection -> state -> state
```

```
isAt :: state -> intersection
headsTo :: state -> intersection

move :: state -> state
turnTo :: intersection -> state -> state
```

Agents are located in this environment at a street intersection and are oriented to move to a neighbouring intersection. They can turn at an intersection to head towards a desired neighbouring intersection and they can move to the intersection they are heading towards. They are modelled after Papert's Turtle Geometry (Abelson and diSessa, 1980; Papert and Sculley, 1980). After a move, the agent heads to the node it came from (Figure 7). This—not quite natural—behaviour leads to the smallest set of axioms for its definition; it can be defined with only four axioms (the operation *connectedIntersections* returns all nodes connected to the node given as an argument):

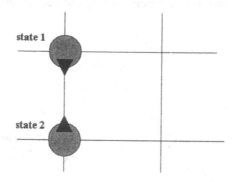

Figure 7: The position of an agent before (state1) and after a move (state2)

1. Turning does not affect the position:

$$isAt\ (turnTo(n,e)) = isAt\ (e) \qquad (2)$$

2. Moving brings agent to the node that was its destination:

$$isAt\ (move\ (e)) = headsTo\ (e) \qquad (3)$$

3. The destination after a move is the location the agent was at before the move:

$$headsTo\ (move\ (e)) = isAt\ (e) \qquad (4)$$

4. Turning (changeDestination) makes the agent's destination the desired intersection:

$$headsTo\ (turnTo\ (n,\ e)) =$$
$$\quad if\ n\ elementOf\ (connectedIntersections\ (isAt\ (e),\ e))\ then\ n \qquad (5)$$
$$\quad else\ error\ ("not\ a\ connected\ intersection")$$

This model is the model of the actions a driver can take on an 'intersection by intersection' level and which are checked against the available street segments (Timpf et al., 1992); drivers are restricted to advance along existing streets. The implementation of the algebra as part of an agent system (for details see Frank, 2000) together with the street network data checks the legality of all moves and calculates the result of such actions. It is a model of a physical agent moving in a street network and is not intended as a model of the human decision process. I call it therefore *basic driving agent*.

4.3 Types of instructions

The algebra of the agent defines the instructions this agent can follow. Instructions are here understood as messages which translate 'piece by piece' into actions. Route descriptions are presented as sequences of instructions, each containing an action word, which translates to an operation, and the appropriate parameters for this action. The algebra with the axioms gives the semantics of instructions and defines which instructions are meaningful for a given agent.

For example, the basic driving agent requires the following instructions to drive from Borders (intersection 1) to Playa Azul (intersection 9) (see Figure 6):

```
startAt 1, turnTo 2, move, turnTo 4, move, turnTo 7,
move, turnTo 9, move
```

All instructions which are meaningful for an agent (defined as an algebra) are of the same type. Typically, all instructions prepared by one web service are of the same type; some web services offer two different instruction types—often including sketches of the intersections in the more detailed one. The four initial route descriptions are all of different types and it is therefore difficult to compare them.

4.4 Instruction equivalence is path equivalence

The result of carrying out a sequence of instructions for driving between two locations is that the agent has travelled through certain street segments and has arrived at the goal location. The instructions describe the path through the network. Two sequences of instructions are equivalent if they describe the same path through the network.

A path is a sequence of locations, starting with the initial location and listing all the locations a driver passes through. Two paths are equivalent if they contain the same location in the same order. Route descriptions of different types can be path equivalent; when carried out, result in the same path.

Equivalence of messages is defined as homomorphism between the algebras of the receivers; it is a well-known fact that homomorphism between algebras establishes equivalence classes (Loeckx et al., 1996). All messages in the same equivalence class define the same pragmatic information.

5   DIFFERENCES IN AGENTS MODELED AS DIFFERENT ALGEBRAS

The instructions given by my friend and the instructions downloaded from the Web do not consist of instructions to move from one location to the next one, as suggested by the 'basic driving algebra'.

For example, my friend assumes that I am able to carry out the operations:

```
followRoadTo :: location -> state -> state

turnTowards :: left/right -> location -> state -> state

followRoadThrough :: location -> state -> state

cross :: streetId -> state -> state
```

This assumes substantial commonsense reasoning, reading and interpretation of street signs; if street signs with the location names indicated are not present, I will have difficulties following the instructions.

For example, the 'basic driving algebra' of moving from location to location can only be used by a person knowing the locations which are mentioned and is clearly not realistic for route information giving. Other methods for giving driving instructions rely on street names (Table 2), on location names on signs on intersections and most use turn directions.

To each instruction (type) belongs a corresponding algebra which explains how to follow these instructions. Trivially, such an algebra contains an operation 'follow one instruction line' with the data in an instruction line as arguments. In this section, different algebras, which each represent a different decision environment, are formalized.

For the following examples, instructions for a path from Borders (Intersection Canon Perdido St and State St, #1) to Playa Azul (Intersection of Santa Barbara St with Cota St, #9) are used (Figure 8). A human could give the following 'natural' route description:

1. Follow Canon Perdido Street to the East for one block
2. Turn right and follow Anacapa Street for two blocks
3. Follow Cota Street to the East for one block

Which results in the path

```
[Intersection 1, Intersection 2, Intersection 4,
  Intersection 7, Intersection 9]
```

5.1   Driver "Turn and move"

In regular instructions using the Basic Driving Agent every turnTo instruction is followed by a move instruction. Merging the two to a single instruction gives ('.' is the composition operation for actions, 'a . b' means do b then a):

```
turnToAndMove :: intersection -> state -> state

turnToAndMove n = move.headsTo n
```

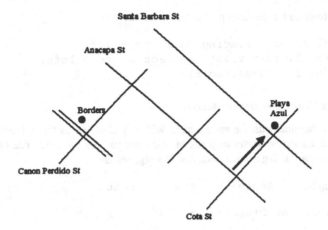

Figure 8: Sketch for path from Borders to Playa Azul

The instruction for the path in the initial example is now:

```
initialize at 1, turnToAndMove 2, turnToAndMove 4,
  turnToAndMove 7, turnToAndMove 9, -> (reached 9)
```

Information content for such a description of a path, not including the information about the start node, is per segment travelled an information about the turn.

5.2 Driver "Turn left/right and move"

A driver who responds to instructions to turn left or right and then move for one segment is using the algebra:

```
turnAndMove :: LeftOrRight -> state -> state
```

The instruction for the same path is for such a driver:

```
initialize at 1 heading to 4, turnAndMove left,
  turnAndMove right, turnAndMove straight,
  turnAndMove left -> (reached 9)
```

5.3 Driver "Turn left/right and move straight for n segments"

A driver who responds to instructions to turn left or right or to proceed for a number of segments is using an algebra like:

```
moveFor :: Integer -> state -> state

turn :: Left_Right -> state -> state
```

The instruction for the same path is for such a driver:

```
initialize at 1 heading to 4, turn left,
 moveFor 1, turn right, moveFor 2, turn left,
 moveFor 1 -> (reached 9)
```

5.4 Driver "Turn and move distance"

A driver not familiar with the environment will pay attention to the indications of the distance and use the odometer to check his movements. He can determine the cardinal directions, perhaps using a small compass. His algebra is

```
turnAngle :: Angle -> state -> state

moveDistance: Distance -> state -> state
```

5.5 Driver "Turn and move till"

This driver is familiar with the environment; in particular he recognizes some street names and is able to read other street names from the signs often found. His algebra is:

```
turn :: Left_Right -> state -> state

moveTill: Streetname -> Turn :: Left_Right -> state ->
 state
```

The interpretation by the driver "Turn and move till" requires information, which is either known to him—"information in the head" in the terminology of Donald Norman (1988)—or information he perceives from the environment—"information in the world."

5.6 Equivalence of instructions

A set of instructions is equivalent if they result in an equivalent path, i.e., when an agent following the instructions touches on the same locations in the same order. This can be tested with simulated execution of the instructions against a representation of the street network (e.g. on a map).

Alternatively, we can translate the different types of instructions listed above into operations of the basic driving algebra (an example was given in 5.1).

The instruction in subsection 5.1 translates to a sequence of instructions for the basic driving agent:

```
Start at 1

turnTo 2, move

turnTo 4, move

turnTo 1, move
```

```
turnTo 9, move
```

which is exactly the instruction given in subsection 4.3. The messages in subsections 5.1 and 4.3 are therefore pragmatically equivalent.

This translation was purely formal and did not require additional information. Others need information from the street network—for example, to translate left or right turns into 'headsTo intersectionID' operations or to translate moveDistance in simple moves along street segments from intersection to intersection.

5.7 Conclusion

The algebra which represents a decision situation defines the method for how information is used pragmatically. Different decision makers with different knowledge encounter different decision situations, i.e. use different algebras for their decision. Instructions for them must be adapted to their knowledge and ability, the instructions must relate to the algebra which describes the decision context; the instructions must use the operations and their parameters according to this algebra.

6 PRAGMATIC INFORMATION CONTENT

6.1 Determination of pragmatic information content

The information content in an instruction of a given type follows from the algebra:
The information content in an action

```
op :: param1 -> param2 -> state -> state
```

is estimated as

$$H = ld(cardinality\ domain\ param1) + ld(cardinality\ domain\ param2) \qquad (6)$$

To this, we have to add the information to select this operation from all the operations in the algebra ($H_o = ld(number\ of\ operations\ in\ algebra)$). There is very often only one operation and therefore H_o is 0 ($ld\ 1 = 0$).

This measure assumes that all combinations of input values are of equal probability (and none illegal—i.e., the function is a total function); if only for some values a valid state change is defined, then the information content is less and must be computed using the formula for entropy (in Section 2).

6.2 Property 1: Different message, same information

A particular agent with a determined algebra expects instructions in the corresponding form. Most humans are versatile and can follow instructions of various types. The algebra of such a decision situation contains the 'basic driving operations' plus some additional ones, which this agent knows how to translate into the basic operations.

The size of the instructions an agent can use varies (see Section 6.1) and if the agent can respond to a number of instruction types, these form equivalence classes of instructions leading to the same actions.

The pragmatic information content for all equivalent instructions an agent in a given situation can use must be the same. Therefore, the information content is the size of smallest instructions in this equivalence class, i.e., the instructions which contain no redundancy (with respect to this agent definition). The beneficial effects of redundancy are not considered in this paper and the question is left for future work.

> The pragmatic information content is the size of the instruction without redundancy for this agent algebra.

Different messages this agent understands may have different data size, but have the same pragmatic information content, namely the data size of the smallest message. This measure is completely dependent on the abilities and knowledge of the agent (modelled as an algebra).

6.3   Property 2: Same message, different information

The same message used by two different agents with different decision context may lead to very different assessment of the pragmatic information content of the message. Compare the agent above which intends to drive, with another agent, which wants only to know when to leave home and therefore needs to know the expected duration of the trip.

```
lengthOfDrive :: [dist&dirInstructions] -> lengthOfDrive
```

For this agent, a specialized message which contains only the expected driving time is pragmatically equivalent with a set of instructions, which contain the distance, which he divides by the expected average speed to calculate the driving time. The pragmatic information content is therefore $ld\ 120 = 7$ bits, for an assumed driving time between 5 minutes and 2 hours.

7   INFORMATION BUSINESS

In this section I sketch how the theory developed here can be used to advance the information business, in particular the business with Geoinformation. In many decision situations, spatial and geographical information plays a role; it is often estimated that 80% of all decisions are influenced by or influence space. In ongoing research we develop methods to assess the value of geographic information in different decision situations as the contribution it makes to improve the decision (Krek, 2002); the assessment of information value is using the same algebraic concepts to model the decision situation as described here.

1. The description of the decision context as an algebra is first helpful for the design of the presentation of results and explanations for the user on how to interpret a route description. The ones found on the Web leave considerable guesswork to the intended use. The pragmatic value of the information is therefore greatly reduced and the user will not trust information difficult to interpret.

2. The measure of pragmatic information content can be used to determine the charges for instructions provided, identifying what is information and what is redundant. For different street network parts (in town, highway, local streets between small towns) different information is necessary for navigation and what is redundant is not always the same.

3. Differential pricing is a key for an effective information business. For uses of information in decision situations which have a higher value, higher prices should be charged, but users will tend to buy information designed for other, lower value uses if they contain all necessary detail.

   For example: If one user must sketch a path for somebody in a map-like way, then instructions with cardinal directions and distances are very useful and other forms of instructions cannot be used. This user takes full advantage of the rich content and deduces higher value from the data. Another user who just uses these instructions to follow a route in familiar territory would translate the instructions in turn and move n segments, and extract only much less information. For example, most of the metric data is just redundant when one moves actually in the physical street network, which keeps drivers on the prescribed roadways. To avoid cannibalism—i.e., that high value users buy the data intended for low value applications—the route descriptions for driving should contain only very approximated cardinal directions and distances, whereas a higher value instruction for drawing sketches of path contains cardinal directions and distances with sufficient precision for the task.

8 SUMMARY

8.1 Pragmatic information content is determined with respect to a decision context

The theory of Shannon and Weaver defines a size measure for the transmission of data; pragmatic information content defines a measure for the amount of information used in a decision context. Two messages are pragmatically equivalent—in a determined decision context—when the decision taken is the same. A decision situation is modelled as an algebra, where the details of the message lead to a decision.

The information content in an action

```
a :: param1 -> param2 -> state -> state
```

is estimated as

$$H\ ld = ld(cardinality\ domain\ param1) + ld(cardinality\ domain\ param2) + H_o \quad (7)$$

where H_o is the information content to select this operation from all possible operations; $H_o = ld(number\ of\ operations)$.

The pragmatic information content for a given decision situation and user is the least amount of data necessary to make the decision. If instructions contain more data, this is redundant, for example, because it is already known by the agent or extracted by him from the real situation. If the same message is used in different decision contexts, then the above method, using a different action for one and the other context, results in different pragmatic information content

8.2 Semantics of instructions defined by model of human user

The semantics of instructions is defined by the decision context, which is a model of the human user. Agents are models of human users of information and can be modelled using algebra. The algebra defines what instructions lead to the same decisions (i.e., what instructions are equivalent). This article concentrates on the general principle of measuring the pragmatic information content and the decision contexts are used only for illustration.

8.3 Open questions

In this article, the algebras were selected for simplicity. It is an important task to determine what good models of human drivers are: what are the abilities of drivers to follow route descriptions? Route descriptions given in natural situations, are quite different from the route descriptions listed in this article initially. Route descriptions produced by humans contain more landmarks:

> Drive down Reinprechtsdorferstrasse till the bright blue coloured store front of the Gazelle chain store;
> Drive along the Taborstrasse till you pass the church;
> Etc.

Messages which are larger than the minimum required for pragmatic actions contain redundancy. This is useful to guard against transmission errors, but also necessary when carrying out the instructions to cope with errors in the instructions and missing information in the world. The assessment of the value of redundancy is an important question, left for future investigations. Small differences between pragmatic information content are certainly overshadowed by the contribution produced from redundant data in unexpected situations.

Different strategies of giving and following route descriptions react differently to errors:

- some fail completely if a minimal error in the instruction is encountered; example: list of turns—one error and a completely different path results which does not lead to the destination.
- some rely on the receiver picking up some additional information from the world; example: relying on street names posted at each corner—fails if these signs are missing.
- some rely on the receiver having specific knowledge of the world.

ACKNOWLEDGMENTS

Support from the Cost Action project and Chorochronos project, both financed by the European Commission, and a project on formal ontology for land registration systems (financed by the Austrian Science Fund) are gratefully acknowledged. The discussions with my colleagues during a meeting in Manchester organized by Michael Worboys were very beneficial. I appreciated the careful review of the article by Christine Rottenbacher, who helped me to see the focus of the paper better.

REFERENCES

Abelson, H. and diSessa, A. A. (1980). *Turtle Geometry: The Computer as a Medium for Exploring Mathematics.* Cambridge, MA: MIT Press.

Birkhoff, G. and Lipson, J. D. (1970). Heterogeneous algebras. *Journal of Combinatorial Theory*, 8:115–133.

Ferber, J. (Ed) (1998). *Multi-Agent Systems—An Introduction to Distributed Artificial Intelligence.* Addison-Wesley.

Frank, A. U. (2000). Communication with maps: A formalized model. In Freksa, C., Brauer, W., Habel, C., and Wender, K. F. (Eds), *Spatial Cognition II (Int. Workshop on Maps and Diagrammatical Representations of the Environment, Hamburg)*, volume 1849 of *Lecture Notes in Artificial Intelligence*, pp. 80–99. Berlin: Springer-Verlag.

Frank, A. U. (2001a). The rationality of epistemology and the rationality of ontology. In Smith, B. and Brogaard, B. (Eds), *Rationality and Irrrationality, Proceedings of the 23rd International Ludwig Wittgenstein Symposium*, Vienna. Hölder-Pichler-Tempsky.

Frank, A. U. (2001b). Spiele als Algebra. In Haller, R. and Puhl, K. (Eds), *Proceedings of the Wittgenstein and the Future of Philosophy, Proceedings of the 24th Int. Wittgenstein Symposium*, Kirchberg a. Wechsel. Austrian Ludwig Wittgenstein Society.

Frank, A. U. (2001c). Tiers of ontology and consistency constraints in geographic information systems. *International Journal of Geographical Information Science*, 15(7):667–678.

Frank, A. U. (2002). Ontology for spatio-temporal databases. In Sellis, T. (Ed), *Spatiotemporal Databases: The Chorochronos Approach*, Lecture Notes in Computer Science. Berlin: Springer-Verlag. To appear.

Krek, A. (2002). *An agent-based model for quantifying the economic value of geographic information.* PhD thesis, Technical University Vienna.

Kuipers, B. and Levitt, T. S. (1978). Navigation and mapping in large-scale space. *AI Magazine*, 9:25–43.

Kuipers, B. and Levitt, T. S. (1990). Navigation and mapping in large-scale space. In Chen, S. (Ed), *Advances in Spatial Reasoning*, volume 2, pp. 207–251. Norwood, NJ: Ablex Publishing Corp.

Loeckx, J., Ehrich, H.-D., and Wolf, M. (1996). *Specification of Abstract Data Types.* Chichester, UK and Stuttgart: John Wiley and B.G. Teubner.

Norman, D. A. (1988). *The Psychology of Everyday Things.* New York: Basic Books.

Papert, S. and Sculley, J. (1980). *Mindstorms: Children, Computers and Powerful Ideas.* New York: Basic Books.

Rao, A. S. (1996). BDI agents speak out in a logical computable language. In van der Velde, W. and Perram, J. W. (Eds), *Agents Breaking Away: Proceedings of the Seventh European Workshop on Modelling Autonomous Agents in a Multi-Agent World*, volume 1038 of *Lecture Notes in Artificial Intelligence*, pp. 42–55. Berlin: Springer-Verlag.

Shannon, C. E. and Weaver, W. (1949). *The Mathematical Theory of Communication.* Urbana, Illinois: The University of Illinois Press.

Timpf, S., Volta, G. S., Pollock, D. W., and Egenhofer, M. J. (1992). A conceptual model of wayfinding using multiple levels of abstractions. In Frank, A. U., Campari, I., and Formentini, U. (Eds), *Theories and Methods of Spatio-Temporal Reasoning in Geographic Space*, volume 639 of *Lecture Notes in Computer Science*, pp. 348–367.

Berlin: Springer-Verlag.

von Neumann, J. and Morgenstern, O. (1944). *Theory of Games and Economic Behavior*. Princeton, NJ: Princeton University Press.

Weiss, G. (1999). *Multi-Agent Systems: A Modern Approach to Distributed Artificial Intelligence*. Cambridge, MA: The MIT Press.

CHAPTER 5

Representational Commitment in Maps

Christopher Habel
Department for Informatics
University of Hamburg, Hamburg, 22527, Germany

1 MAPS: EXTERNAL REPRESENTATIONS FOR COMMUNICATING SPATIAL INFORMATION

Maps are representational means that enable people to solve different types of problems with respect to the external, geographical world. From the perspective of the map user, the map is a knowledge source, which has to be interpreted, to become applicable as background knowledge for solving spatial or geographical problems. Interpretation—as a cognitive process—depends on the user and the context. In other words, different users will come to different results using the same map in the same problem-solving task, due to their subjective differences in map interpretation. From the perspective of the mapmaker, the mapping from a region of the external, geographical world to the representation, i.e. to the map, should follow principles that guarantee some objective conditions of correspondence between the representation and the represented part of the world. Following this line of thought, cartography can be characterized as the scientific discipline that investigates the principles of correspondence on the basis of mathematics as well as empirical sciences. On the other hand, the mapmaker has to produce a map that is interpretable by the map user for solving spatial problems. Taking this view of "maps as representations for communication" makes ensuring the success of communication a prominent task in the production of maps (see MacEachren, 1995 and Frank, 2000 on the communicational function of maps).

To sum up, if people solve spatial or geographical problems by using maps, map producers are implicitly involved in this process as knowledge brokers who transact information about the external, geographical world to the map user (see Figure 1).

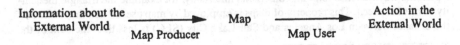

Figure 1: The map as mediating external representation

Today, in the time of GIS, a relevant type of map production is performed on the fly: Automatic visualisation systems create sketch maps task-dependently for individual spatial problems.[1] In traditional processes of map production, the science of cartography provides well-founded principles of correspondence between the represented domain and the map, i.e. the representation. In contrast, visualising geographic information in a sketch map belongs to the class of ill-defined tasks in the sense of Simon (1980), i.e. success in solving the sketch map production task can be evaluated with respect to a variety of criteria leading to criteria-dependent, graded valuations. There is a strong and fascinating correspondence between Simon's (1980) discussion of investigating ill-formed tasks, as natural language comprehension in Artificial Intelligence, and the analysis of the state of the art and the research themes published in the research agenda of the ICA's Commission on Visualization (Slocum et al., 2001).

In this article I suggest a semantic-pragmatic approach in the tradition of systematic pragmatics—as proposed by Grice (1989) and Levinson (2000)—to characterise principles for producing sketch maps suitable to *successful communication*. In other words, these principles specify properties of sketch maps that ensure that the maps can be interpreted easily and unambiguously. The formal basis of the set principles are spatial, i.e. topological and geometric, properties of maps, seen as external representations.

1.1   A short outline of representation theory

External representations play a major role in human cognition and especially in human problem solving (Zhang, 1997). Two types of external representations are in the focus of cognitive science research: On the one hand, propositional, sentential representations—often characterized as *linguistic* or *language oriented*, which contain language as well as number symbols—on the other hand, analog representations, also called *pictorial, diagrammatical* or *image-like*, e.g, maps, diagrams, etc. The benefits of analog representations[2]—whether they are internal or external, whether they are called pictorial, spatial or diagrammatic—are based on their property of sharing relevant inherent constraints with the domain they represent. In contrast, propositional representations must code such constraints, which are implicitly embodied in analog representations, explicitly (see Palmer, 1978, p. 271; Haugeland, 1987, pp. 88 ff.; Myers and Konolige, 1995, p. 277).

I will exemplify this idea of inherent constraints with a comparison between mileage charts, which are two-dimensional propositional representations (see Figure 2a), and distance maps, such as Figure 2b which is a pictorial representation containing the information printed in boldface in Figure 2a. The map, which represents spatial configurations in a pictorial way, satisfies relevant constraints inherently, for example the triangle inequality of metric spaces. This property of maps is employed by map users when they infer that the distance between Los Angeles and Salt Lake City (SLC) is less than the sum of the

---

[1] In the present article I use the term *sketch map* for graphical representations of the environment produced by layperson, where "lay" refers to lack of cartographic expertise. I cover under this term hand-drawn maps as well as maps produced using graphical software tools. Thus, sketch maps produced by advertising agencies can also fall under this notion.

[2] Strictly speaking, analog representations are analog with respect to specific aspects of the domain to be represented. The assumed structural correspondence between representation and represented domain holds with respect to some prominent properties and relations (see Palmer, 1978).

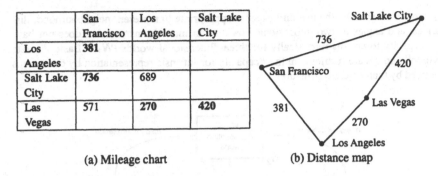

	San Francisco	Los Angeles	Salt Lake City
Los Angeles	381		
Salt Lake City	736	689	
Las Vegas	571	270	420

(a) Mileage chart (b) Distance map

Figure 2: Mileage chart and distance map for 4 major US cities

distances between LA and Las Vegas (LV), and that between LV and SLC, respectively, even if some distance labels are missing or are incorrect. On the other hand, missing or incorrect data in distance tables have to be checked explicitly. For example, a misprint in a mileage table is more cumbersome to detect than similar errors in a distance map.

From a representational point of view, these are problems with respect to explicit or implicit triggering of inferences. Before I elaborate on this topic in later sections I introduce the main representational concepts to be used in the following. Following Palmer (1978, pp. 263ff.) , *representation constellations* can be characterized by the *represented world* (W_1) and the *representing world* (W_2), often abbreviated as *representations*, their *internal structures*, and a partial mapping ρ from W_1 to W_2, which is called the *representational mapping* (see Figure 3). In a mathematical analysis of a representational constellation internal structures of worlds are described by relational systems. An analog representation of a world, e.g., a map, has to fulfil specific requirements on the representational mapping. Furthermore, to use W_2 successfully as an external representation—especially as means of communication as depicted in Figure 3—it is necessary that people using W_2 know how ρ maps W_1 to W_2. In particular they possess knowledge about correspondences between *world concepts* (*WC*) and *map concepts* (*MC*), e.g., that streets can be represented by lines, that rivers can be depicted by blue lines, etc.[3] Furthermore, for successful communication of spatial information by the means of maps the conceptual system of world concepts and map concepts hold by the map producer, i.e. *WC(mp)* and *MC(mp)*, should be in agreement with that of the map user, *WC(mu)* and *MC(mu)*.

Pictorial representations (W_2) are realized on a medium with spatial characteristics, on which relevant spatial properties and relations are intrinsically given. Drawings on paper or sketch maps presented on a computer display exemplify realization in the case of external representations. Activation patterns in symbolic arrays (Glasgow and Papadias, 1992) or in the visual buffer (Kosslyn, 1980, 1994) exemplify realization in the case of internal representations. The inventory of spatial properties and relations intrinsically realized in the medium determines the representational capacity of the representational

---

[3] People's knowledge about such correspondences is the basis of their competence for map using. In particular, they have to be aware of different ways of coding, i.e., of conventions for representational mappings. On the other hand, in professional maps the principal correspondences are specified by legends.

system. For example, the medium paper is appropriate to represent neighbourhood, distances, and angles of two-dimensional space in an intrinsic manner, but does not have the capacity to do this intrinsically for three-dimensional worlds (W_1). Basic discrete symbolic arrays are restricted in their capacity for intrinsic representation by constraints induced by their discrete structure.[4]

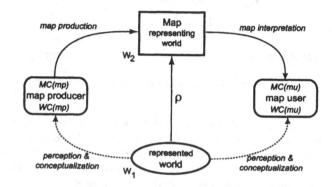

Figure 3: Representation constellation

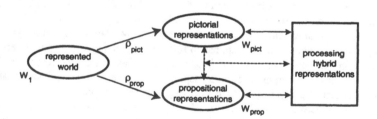

Figure 4: The basic structure of a propositional–pictorial hybrid architecture

I use the term *pictorial elements* for the entities in pictorial representations. In the present paper it is sufficient to distinguish between two kinds of pictorial elements, on the one hand, *basic graphical elements*, as lines, angles, rectangles, etc., and, on the other hand, *complex graphical elements*, which are configurations of basic graphical elements combined to represent specific concepts of the represented world (see Kosslyn, 1989, for a kindred approach for analysing charts and graphs, seen as means for communicating quantitative information). The pictorial elements—in particular the complex ones—can

---

[4]In Habel (1994) I argue for a system of representation systems to overcome some limitations of discrete systems; by such systems of discrete systems it is possible to "approximate" density.

be linked with elements of a propositional representation system. Integrating both kinds of representation in one system leads to representation systems which are called *hybrid* or *heterogeneous* (see Figure 4). Barwise and Etchemendy (1995) propose an implicit linking via mappings between the represented world W_1, and the two types of representations (representing worlds), W_{pict} and W_{prop}. On the other hand, Myers and Konolige (1995), Glasgow and Papadias (1992) as well as Habel (1987) and Habel et al. (1995) propose explicit linking via labeling the pictorial representation W_{pict} with symbolic names referring to W_{prop} or employing "referential links" between W_{pict} and W_{prop}.

A wide spectrum of empirical evidence—from perception and imagery (see, e.g., Kosslyn, 1980), from reasoning (see, e.g., part III of Glasgow et al., 1995) and from language (see, e.g., Landau and Jackendoff, 1993)—supports a widely held hypothesis, which Habel et al. (1995) expressed as follows:

> *Principle of hybrid representation and processing*: A cognitive system that reasons successfully in a general (non-restricted) real world environment is based on a hybrid architecture, which combines propositional and pictorial representations and processes.

By externalising our internal representations of the external world, we are able to transfer our experience to other human beings and to use the experience and expertise of others in problem solving. Whereas people know—mostly—how to use the external representation they produced themselves, e.g. in the case of using paper and pencil for doing arithmetic with external representations, they often have difficulties in interpret representations produced by others. Thus, the two roles of being a producer of representations and being a user of representations have to be considered in detail.

1.2 Communication about space

According to the paradigm of *External-Representation based Problem Solving* (Zhang and Norman, 1994; Zhang, 1997), maps are external representations used—in interaction with internal representations of the map user—to solve problems with spatial or geographic character. As a consequence, the wide variety of spatial and geographic tasks is reflected in various types of maps. In other words, there is a close correspondence between classes of spatial or geographical problems and types of maps.

Traditional maps, such as city maps or state maps, are multi-purpose means of spatial problem solving. For example, for any pair of locations A and B represented in such a map, the map is a helpful means to solve the wayfinding problem from the origin A to the destination B. More specialized sketch maps like those designed for finding the way to a specific shopping mall or a hotel are not entirely determined on an individual wayfinding task: While they are fixed with respect to the destination, they hold information relevant for starting from a (limited) number of different origins. Individually constructed maps stand in contrast to such multi-purpose maps, e.g., hand drawn sketch maps or maps generated by driving-direction systems (as MapQuest) are produced to give assistance in finding the way for one specific origin-destination pair.

Map producers, who create an external representation containing the information needed to solve other people's spatial problems, are involved in a communication task similar to speakers, i.e. language producers, who generate verbal route instructions (compare the schemata depicted in Figures 1, 3). On the other hand, people interpreting a map for wayfinding have to perform comprehension processes similar to those carried out by listeners or readers in understanding verbal route descriptions.

For more than twenty years, route descriptions have been subject to interdisciplinary research in linguistics and psychology (Klein, 1982; Wunderlich and Reinelt, 1982; Allen, 1997; Denis, 1997; Tschander et al., 2002). These investigations focus on the communicative roles of the instructor and the instructee, as well as on the processes of producing and comprehending route descriptions. Additionally, there are also specific studies on criteria for 'good' route descriptions (Lovelace et al., 1999), on the influence of types of environment on route descriptions (Fontaine and Denis, 1999), and on the correspondence between structures of route depictions and structures of route descriptions (Tversky and Lee, 1999).

Due to the correspondence between processing maps for wayfinding and verbal route descriptions, it is reasonable to investigate which results from linguistic and psycholinguistic research—in particular, which findings on route descriptions—are reasonable to transfer to the field of 'maps as external representations for communicating spatial information'. Furthermore, taking the perspective of *communicating by maps* urges us to take a closer look at the general principles of successful communication, as discussed in the Neo-Gricean approach of *systematic pragmatics* of Levinson (2000).[5] Grice's program for a theory of—primarily verbal—communication (see Grice, 1975, 1989) assumes two levels of meaning, namely *sentence-meaning*, i.e. what the sentence 'means' independently of context, and *utterer's-meaning*, i.e. what the utterer intends to communicate. An additional core concept of systematic pragmatics is that of *implicatures*. These are non-logical inferences providing additional content beyond what is said explicitly with a sentence. From a language comprehension perspective, implicatures are performed for computing utterer's-meaning from sentence-meaning. Thus, knowledge about implicatures and how to execute them is central for linguistic communication. Levinson focuses on a third, intermediate level, namely that of *utterance-type-meaning*, which can be seen as the "level of systematic pragmatic inference based on ... general expectations how language is normally used" (see Levinson, 2000, p. 22).

In the following section, I explain some core concepts of systematic pragmatics in more detail exemplified with two types of sketch maps. By this, I indicate how Grice's and Levinson's ideas can be used to work out some 'communication principles of good sketch maps', kindred to the perceptual and cognitive principles Kosslyn (1989) proposed for charts and graphs. These principles of visualising geographic information for casual users are twofold. Firstly, they are fundamental in interpreting sketch maps, e.g., for using this type of external pictorial representations successfully in wayfinding. Secondly, the principles are guidelines for the mapmaker to produce well-designed maps easy to use in spatial problem solving. In Section 3, I propose a formal approach to describe principles for representational commitment in maps. These principles can be used as guiding principles for designing visualisation devices as well as for the evaluation of appropriate design of sketch maps.

---

[5]Oberlander (1996) and Gurr (1999) discuss kindred views for other types of pictorial representations, e.g., network diagrams, Venn diagrams, Euler's Circles, etc.

2 COMMUNICATING SPATIAL INFORMATION BY MAPS: COMMITMENT AND IMPLICATURES

2.1 Problems of determination in pictorial representations

According to Palmer's approach to representation theory (see Section 1.1), maps belong to the class of analog representations that are realized graphically, or in other words, pictorially. The *analog character* of maps is due to structural constraints, which have to be satisfied by the representational mapping ρ. Actually, ρ is a homomorphism that preserves the relevant structural properties of the represented world (see Worboys, this volume, who presents a kindred, information-theoretic approach based on the notion of *infomorphism*).[6]

The analog character of graphical representations does not only cause its major benefits in problem solving (see Section 1.1), but is also the origin of some disadvantages. From a representational point of view, these are especially problems with respect to commitment to details and determination, which subsumes the question of accuracy and precision discussed by Worboys (this volume). These problems are ubiquitous with pictorial representations: if you use paper and pencil for producing a sketch map (such as depicted in Figure 5a) or do it using a graphical tool on your computer (as in Figure 5c), then the medium can bring about determination not intended by you. For example, since it is not possible to draw a T-junction without giving it a determined angle between the streets, the realized angles—76° and 104° in Figure 5c–d—can be artefacts of drawing. Furthermore, the bends in the lines of Figure 5b (marked by arrows) can be due to inaccuracies in the drawing process as well as they can be intended to represent bends in the real world. In contrast, similar bends or kinks in a diagram produced with a graphical tool (as in Figure 5e) will be usually interpreted as representing real world bends.[7]

Commitment to detail and determination is a core problem of pictorial representations. To give some examples: Myers and Konolige (1995) discuss this topic using the term *structural uncertainty*, and Ioerger's (1994) proposal of a method of using images as spatial representations is presented as a solution of the *indeterminacy problem*. Since Ioerger's argumentation reflects some of the presuppositions of anti-pictorialists, it is informative to have a closer examination of his approach.[8] With respect to the indeterminacy problem Ioerger formulates some basic assumptions, namely, (i) "images are by definition determinate; as models they entail specific answers to every query." (p. 559); (ii) "The Indeterminacy Problem stems from the fact that an image can only represent a single model." (p. 561); (iii) "once an image is constructed, it becomes impossible

---

[6]Note, that Palmer (1978) characterises the representational mapping as an isomorphism instead of an homomorphism. This seems to be caused by an additional, non-explicit preparatory step of focussing firstly on the relevant structural properties, which precedes the actual representation mapping.

[7]This interpretation is based on the map users' assumptions about the mapmaking process. Bends in hand drawings are common, thus they are often not seen as meaningful. On the other hand, in using a graphical tool the production of a line sequence consisting of three segments (Figure 5e) requires more effort than the drawing of one line (Figure 5c). This justifies the assumption that the kinks are intended to represent. In performing this way, map users follow the *Principle of Communicative Nonaccidentalness* I discuss in more detail in section 4.1

[8]Anti-pictorialists are one side in the so-called 'imagery debate', a long-hold controversy about the status of mental images, see Tye (1991); Glasgow (1993). The view that pictorial representations are fully committed is widely held in the Artificial Intelligence community and can even be found with researchers on Diagrammatic Reasoning.

Figure 5: Depicting T-junctions

to tell the difference between a detail that was implied by the assertions from a detail that was arbitrarily fixed during image construction." (p. 561). To solve the indeterminacy problem Ioerger proposes a method kindred to the conception of *multiple-models* developed by Johnson-Laird and Byrne (1991), which they call ISR (≈ indeterminacy in spatial reasoning). ISR "avoids the indeterminacy problem by considering multiple images to discover which details are implied and which are arbitrary" (p.562). I will sum up this idea for solving indeterminacy as: (iv) 'indeterminacy is represented and processed by disjunctions of fully determined representations'. In contrast to the above described pseudosolutions of the (in-)determinacy problem, I describe in the following an alternative approach based on the concept of *commitment*, which allows us to treat spatial and diagrammatic representations as underdetermined.

2.2   Commitments in interpreting sketch maps: Two example domains

To exemplify the role of commitment in interpreting sketch maps I use floor plans similar to the hallway maps of Myers and Konolige (1995). Whereas they discuss this domain focusing on the spatial knowledge a mobile robot might contain, I deal with floor plans as sketchy pictorial representations of spatial layouts, which might be given to a human in the form of a sketch map, i.e. an external representation.

Figure 6 presents two floor plans depicting the location of office F-32. In interpreting plan (a) users have two options: (1) they can assume that there are four rooms on the 3rd floor, namely three at one side and one large room at the opposite side, (2) they can see the rectangles not labelled by a room number as white spots on the floor plan giving no information about the rooms in the corresponding areas of the floor. Since the floor layout resulting from option (1) is—for most types of buildings—unusual, there is—without additional situational context—a tendency to choose alternative (2). In contrast, map (b) holds the information that there are 6 rooms at each side of the corridor. Thus, if a map user would interpret plan (a) assuming the existence of only 4 rooms on the 3rd floor we would be justified to see this as a misinterpretation on the behalf of the interpreter. On the other hand, if in reality there were not six offices at each side of the corridor, the misinterpretation about the number of offices on the 3rd floor would be due to the producer of plan (b).

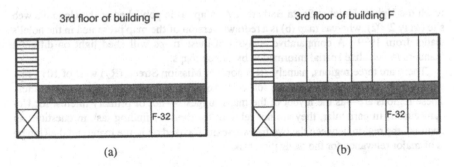

Figure 6: Floor plans depicting the location of office F-32

Using the idea of producers' commitments, the difference between the sketch maps (a) and (b) can be described as follows: floor plan (a) is committed to the existence of office F-32, and is not committed to the existence of other specific rooms on the 3rd floor. In contrast, floor plan (b) is committed to the existence of six offices at each side of the corridor. The two types of commitment exemplified here, I call in the following *commitment to existence* and *cardinality commitment*. Only if map producer and map interpreter take these commitments into consideration, misinterpretation of a map can be avoided. Or, in other words, to consider the commitments inherent in a sketch map is the basis for using them successfully in spatial problem solving. For example, since map (b) is committed to cardinality, people using this map are justified to identify F-32 as the fifth office at the right side of the corridor (seen or coming from the staircase).

Now let me return to the general conception of the representational mapping ρ (cf. Figure 3). *Commitment to existence* concerns inferring from the existence of entities in the map to existence of entities in reality. These inferences are based on correspondences between map concepts and world concepts. I exemplify this with the case of floor plan (a). A standard correspondence for sketch maps depicting floors is that rooms are represented by rectangles. The rectangle labelled with F-32 is an instance of this type, to be precise: the room named F-32 is mapped by ρ to the rectangle labelled with F-32. Therefore it is possible to infer, that other rectangles—of similar character—are also the images of applying ρ to other rooms. Thus, the large rectangle (above the shaded rectangle) could be the representation of a room. The white spot interpretation, which holds no commitment to existence, suspends this inference. In this case, the real world counterpart of the large rectangle can also be a *group of rooms*, i.e. an instance of another concept. *Cardinality commitment* leads to a stronger inference concerning the cardinality of entities: if there are exactly n entities of a specific type in the map, e.g., room representing entities, then there exist exactly n entities of the corresponding type—namely rooms—in the world.[9]

Some additional problems of determination and commitment will now be exemplified with the maps depicted in Figure 7, which are produced to give customers information to

---

[9]To know which map concepts can hold cardinality commitments and which map concepts (nearly) never hold these commitments is an essential part of our competence for map using. For example, the small rectangles in the upper left corner of the maps depicted in Figure 6 do not represent individual steps, but they function as pattern to represent stairs.

reach the Upham Hotel at Santa Barbara, CA. Map (a) is published at the Upham's web site (May 2002), whereas map (b) is a redrawn version of the map presented in the hotel's leaflet from 1999. A comparative analysis of these maps will shed light on different manners to visualize spatial information by sketch maps.

There are three regions, namely, (R_1) north of Mission Street, (R_2) west of 101 Freeway, and (R_3) south of 101 Freeway, for which both maps do not contain any information. These regions are—as the layout of the maps suggests—not of primary interest for Upham guests, in particular, they are not relevant for the way-finding task in question. In contrast, the region between the two freeway-exits focused on in the maps and the Upham is of major relevance for the navigation task.

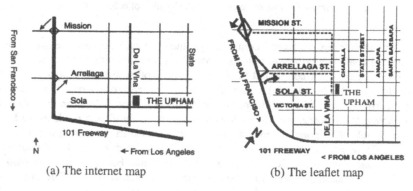

(a) The internet map   (b) The leaflet map

Figure 7: Two sketch maps directing to *The Upham* (Santa Barbara, CA)

Let us consider what information users could try to get from the maps: since counting blocks is—in particular after dark—a very successful heuristic to identify the street to turn into, confidence in information about the ordinal numbers of streets on a route is desirable. If you apply the counting-the-blocks strategy during the final phase of the route—beginning with the Mission St. exit—this would lead to "second right and then in the third block at the left" using the internet map (Figure 7a) and to "second right and then in the sixth block at the left" for the leaflet map (Figure 7b). As the veridical map depicted in Figure 8 illustrates, the real situation differs from these results of map interpretation. Whereas the internet map (Figure 7a) is completely unsuitable for using the counting-the-blocks strategy, the more detailed map (Figure 7b) causes only a minor difficulty, since this map does not inform you about a small road to the right, that you reach immediately before you have to turn into De La Vina (cf. Figure 8). To sum up, the internet map (Figure 7a) is committed to two distinguished starting points, namely the Mission St. exit and the Arrellaga St. exit, as well as to the end point of the route, i.e. to *The Upham* as goal. Thus, the two floor plans depicted in Figure 6 and the two sketch maps directing to *The Upham* (cf. Figure 7) are members of two classes of maps that can be characterized with respect to existence and cardinality commitments: the first class, (e.g., Figures 6a and 7a) is committed only to starting locations and goals of wayfinding

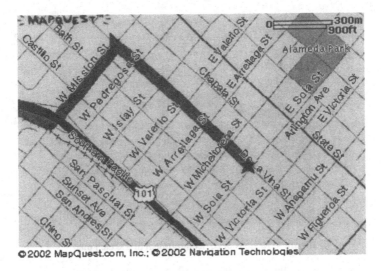

Figure 8: A veridical route map to *The Upham* (map provided by MapQuest.com, Inc.)

problem, whereas the second class (e.g., Figure 6b and 7b) is committed to the existence and cardinality properties induced by the graphical entities in the map. The commitment relevant for the counting-the-blocks strategy is the requirement that the representational mapping is a bijection between the set of streets relevant for the wayfinding task and the set of street-representing lines in the map. This requirement leads to the rule "If there is no street in a specific region of the map, then there is no street in the corresponding area in reality." Applying this rule in using the internet map would lead to failures in wayfinding.

Let us now concentrate on the leaflet map (Figure 7b). The commitment to existence and cardinality is followed in the region between Mission Street and Victoria Street, and between the Freeway and State Street. But, there is another region, namely that south of Victoria Street, which is only partially veridical, i.e. in which the rule of 'bijective correspondence between streets and street-lines in the map' is broken: in the real environment you find nine streets parallel to Victoria Street and three freeway exits south of Victoria Street not depicted in the map.

Regions in which a set of specific commitments holds, e.g., the existence and cardinality commitments induced by the bijection requirement, are important in using maps. In the following I call them *regions of equicommitment* (cf., e.g., Figure 9). Successful communication by maps requires that map users are able to recognize areas in the maps as regions with common properties.[10] They are able to detect such regions, since the conventions of map-making give hints how areas of *equicommitment* are constituted. In particular, in the case of task-specific maps the task to be solved is the functional background for the interpretation process. In our example (cf. Figure 7) the task is to reach *The Upham*. The goal, *The Upham*, and the starting points of the task, the two relevant

[10]This formulation does not mean that map users are conscious of the conception *regions of equicommitment*, but that they use it implicitly.

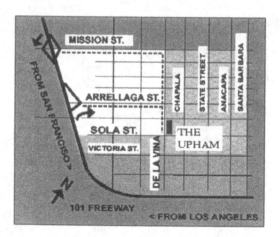

Figure 9: Regions of equicommitment in a sketch map

freeway exits, are distinguished entities. Furthermore, the use of name labels in the map is the other relevant source of information, e.g., the streets parallel to De La Vina are annotated with names, whereas the streets south of Victoria do not have name labels. By use of common, very successful heuristics, namely, that *regions of equicommitment* are often convex and in city maps often rectangular, a classification of *regions of equicommitment* as depicted in Figure 9 results.

To sum up, the successful interpretation of a map depends, on the one hand, on the user's understanding of the commitments, which are coded in the map, and on the other hand, on the producer's ability to make the coding of commitments detectable. In line with *systematic pragmatics*—see, e.g., Grice (1989); Clark (1996); Levinson (2000); cf. section 1.2—I propose in the following section formal methods for specifying representational commitment in maps.

3 SPECIFYING THE GEOMETRICAL STRUCTURE OF MAPS

In understanding maps three levels of representations and analyses are involved, the syntactic level regarding the graphical entities and spatial configurations of graphical entities as well as their detection in visual perception, the semantic level, in which the meaning of graphical entities and their configurations is dealt with, and the pragmatic level, which concerns the content of the map—or rather, the content of focused parts of the map—as it is involved in situations of map interpretation (cf. MacEachren, 1995; Kosslyn, 1989, on understanding graphs). Since the basic perceptual processes of recognising meaningful units in sketch maps are not the topic of the present paper, I presuppose the successful interpretation of the graphical elements. Instead, I will concentrate on the interacting processes of semantic and pragmatic interpretation of maps.

3.1 The hallway domain – revisited

For discussing the indeterminacy and commitment phenomena described above from a geometric perspective, I come back to the hallway domain. The formal characterisation of commitment I propose in this section will be exemplified with a highly schematic map similar to that examined by Myers and Konolige (1995)—see Figure 10—but more elaborated than theirs, to have a richer and therefore more interesting geometrical structure, which will be central for the following (see Figures 11a,b).

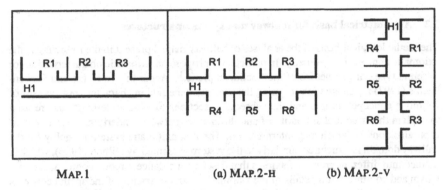

MAP.1   (a) MAP.2-H   (b) MAP.2-V

Figure 10: One-sided hallway map

Figure 11: Two maps of the same hallway

The layouts represent that the spatial constellation contains seven (or four) entities, namely a hallway (with the symbolic name H1) and six (or three) rooms (R1 to R6, R1 to R3 respectively). The interpretation of symbolic names leads on the semantic level to information about the category of the entities depicted in the map, e.g. 'R-labeled entities represent rooms'.

Discussing a hallway map which has the spatial structure of MAP.1, depicted in Figure 10, Myers and Konolige (1995, p. 276) use the natural language description "the leftmost office" for a room in the position of R1. This description is suitable with respect to MAP.1 only if a viewer's perspective is taken, i.e. if the horizontal layout of the floor plan—realized on paper or on a computer screen—is interpreted as a distinguished left-to-right-orientation induced by the standard direction of reading (e.g. in English). But the description is not appropriate to the represented room on the represented floor without the intermediating link of the map.

Beyond the spatial regions representing rooms and the hallway, that are marked by symbolic labels, at least three spatial relations between spatial entities are represented in the maps depicted in Figure 11 (see Table 1).

On the basis of these relations, MAP.2-H andMAP.2-V provide information corresponding to verbal descriptions as the following: "R1 and R3 are next to each other," "R2 is on the same side of the hallway as R3," "R2 and R5 are on different sides of the hallway" or, more specifically, "R2 is opposite to R5."

Relational concept	Meaning	Instances
ADJACENT(x, y)	x is adjacent to y	⟨H1,R1⟩, ⟨H1,R2⟩, ⟨H1,R3⟩, ⟨H1,R4⟩, ⟨H1,R5⟩, ⟨H1,R6⟩, ⟨R1,R2⟩, ⟨R2,R3⟩, ⟨R4,R5⟩, ⟨R5,R6⟩
BETWEEN(x, y, z)	x is between y and z	⟨R2,R1,R3⟩, ⟨R5,R4,R6⟩, ⟨R4,H1,R1⟩, ⟨R2,H1,R5⟩...
OPPOSITE(x, y)	x is opposite to y	⟨R1,R4⟩, ⟨R2,R5⟩, ⟨R3,R6⟩

Table 1: Three spatial relations of the hallway domain (MAP.2-H, MAP.2-V)

3.2 A geometrical basis for hallway maps – linear structures

The methodological basis of the analyses of hallway maps I present in the following is the axiomatic approach to geometry. In particular, I propose an axiomatic characterisation for different types of geometries of the Euclidean plane.[11] According to Habel and Eschenbach's (1997) argumentation for using the axiomatic method in characterizing systems of cognitive concepts, in this and the subsequent sections axiomatic descriptions are used to explain the structural similarity of and differences between underlying cognitive conceptualizations in sketch map interpretation. The first axiomatic system I employ for the hallway domain is structured similarly to the system presented by Hilbert (1956), which is divided into different groups of axioms that describe incidence, order, congruence, parallelism and continuity (on details of the axiomatic characterization of the spatial concepts in question, see Eschenbach and Kulik, 1997, and Eschenbach, 1999).[12] The inventory of geometric entities used here consists of *points*, *(straight) lines* and *oriented lines*, which can be seen as lines supplied with an orientation, but are assumed to be an independent concept. Furthermore, there are two important relational concepts:

- a binary relation of *incidence* (symbolized by ι), whose domain is the set of points and whose range is the union of lines and oriented lines. This relation corresponds to situations of a point lying on a line or an oriented line or to situations of an oriented line going through a point.
- a ternary relation of *precedence* (symbolized by $<$). It combines oriented lines and two points. With this relation, ordering of points on oriented lines can be described.

Two groups of axioms, the axioms of incidence and ordering on linear structures and the definition of *betweenness* are sufficient to characterize the spatial concepts ADJACENT, BETWEEN and OPPOSITE listed in Table 1 (see, Eschenbach and Kulik, 1997, and Eschenbach, 1999).

---

[11]In the present paper I do not tackle the details of the internal topological structure of the plane, i.e. the difference between discrete (finite), dense or continuous 2-dimensional spaces (cf. Habel, 1994, with respect to the temporal domain). For the present objective, namely the discussion of commitment phenomena of pictorial representations, it is possible to pass over these differences.

[12]Eschenbach and Kulik (1997) develop an axiomatic characterization of spatial orderings and the concepts underlying intrinsic and deictic uses of spatial terms such as *in front of, behind, left* and *right*. Eschenbach (1999) proposes a model for the geometric structures of 'spatial reference systems' and a systematic characterization of the terms 'direction', 'perspective' and 'point of view'. In both approaches the geometric concept *betweenness* plays a fundamental role (cf. Tarski, 1959).

Figure 12: Linear structures brought on in the hallyway map during comprehension

Figure 13: Grid graph abstraction of the hallway domain

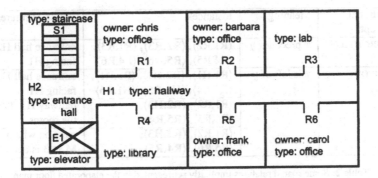

Figure 14: An elaborated hallway map: MAP.3

Linear structures act in two ways in the comprehension of maps. On the one hand, there are linear graphical entities, e.g., lines that depict street as in the Upham maps (see Figure 7), on the other hand, interpreters of maps bring on linear structures during the process of comprehension. For example, hallways whose proportion fulfils specific requirements are conceptualised as linear, as well as the arrangements of rooms on individual sides of such hallways can be seen as linear (see Figure 12). Figure 13 depicts a further abstraction of the hallway map, which I call *grid graph*, whose edges represent rooms or *in-front-of-door locations* on the hallway. This abstraction can be seen as a two-step process: firstly, conceptualising linear structures—as depicted in Figure 12—builds up a grid on the map (on grids as means for qualitative spatial reasoning, see Kulik and Klippel, 1999). Secondly, a discrete version of the grid, the *grid graph*, can be used as an independent representation for spatial problem solving, e.g., the sequence R1, h12, R4 stands for the path to follow passing from Room R1 to the opposite room R4.

To sum up, during interpreting a map linear structures are brought on, that provide an inventory of spatial relations, in particular, BETWEEN, ADJACENT and OPPOSITE. These relations are fundamental in spatial problem solving, e.g. in wayfinding, as well as they are the basis for natural language descriptions of space.

3.3 Extending the geometry of hallway maps – directed and planar structures

In contrast to the maps given in Figure 11 the elaborated floor map, MAP.3 (Figure 14), provides additional information which brings on a more specific geometric structure than that given by MAP.2-H or MAP.2-V, namely one which contains—beyond the spatial concepts discussed in the previous section—additional concepts (see Table 2). This enriched geometric structure allows more specific verbal descriptions, such as *"You reach Barbara's office after you passed Chris' office"*[13], *"The lab is on the left side"* or *"Chris' office is left of Barbara's office"*. In particular, this additional structure is useful for spatial problem solving, e.g. using the *counting-the-rooms-passed strategy* to find a specific goal.

Relational concepts	Meaning	Instances	Frame of reference
IN_FRONT_OF (x,y)	x precedes y	$\langle R1,R2 \rangle, \langle R2,R3 \rangle, \langle R1,R3 \rangle,$ $\langle R4,R5 \rangle, \langle R5,R6 \rangle, \langle R4,R6 \rangle$	entrance hall H2: facing H1
LEFT_OF (x,y)	x is left of y	$\langle R1,H1 \rangle, \langle R2,H1 \rangle, \langle R3,H1 \rangle,$ $\langle H1,R4 \rangle, \langle H1,R5 \rangle, \langle H1,R6 \rangle$	entrance hall H2: facing H1
		$\langle R1,R2 \rangle, \langle R2,H3 \rangle,$ $\langle R6,R5 \rangle, \langle R5,R4 \rangle$	hallway H1: facing the rooms
		$\langle R1,R2 \rangle, \langle R2,H3 \rangle,$ $\langle R5,R6 \rangle, \langle R4,R5 \rangle$	page on which MAP.3 is depicted

Table 2: Some spatial relations implicitly represented in the elaborated floor map

The meaning of directional spatial concepts, as IN_FRONT_OF and BEHIND as well as LEFT_OF and RIGHT_OF,[14] depends on the *frame of reference* selected for interpretation, in particular, the choice between an *intrinsic*, an *absolute* or a *relative* frame of reference can be crucial in understanding a verbal description (see Levinson, 1996; Eschenbach, 1999). For example, the two cases of "left" mentioned in the verbal descriptions, differ in the type of reference frame determining the interpretation: in *"The lab is on the left side"* the absolute frame brought on by the configuration of entrance hall and hallway is used. Processing the phrase *"left of Barbara's office"* applies either a relative frame of reference provided by an imagined person facing the rooms from the hallway, or refers to the absolute frame built up by the page on which MAP.3 is printed; note, that describing Carol's office with respect to Frank's office would lead to different descriptions.

What leads the map user to seeing additional structure in MAP.3 is due to a conventional perspective based on experience about moving in buildings: if a hallway possesses one entrance—as it is the case in the floor depicted by Figure 14—then this distinguished area of the hallway is regarded as the origin. In other words, this perspective induces direction. From a geometrical point of view, the hallway brings on a *directed line*—to be precise, a line segment—H1, with a distinguished starting point $st.p(H1)$, see Figure 15.

---

[13] In German exists a purely spatial variant of this description: *"Barbaras Büro is direkt hinter Chris Büro"*. (*"Barbara's office is immediately behind Chris' office"*).

[14] IN_FRONT_OF and BEHIND as well as LEFT_OF and RIGHT_OF are pairs of concepts converse to each other (for details on *converseness*, see Eschenbach and Kulik, 1997). Thus, I skipped the instances of the converse relation in Table 2, and will discuss in the following only one member from each pair of relations.

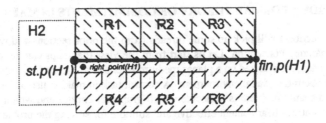

Figure 15: Geometrical structure underlying the elaborated hallway map

IN_FRONT_OF, i.e. precedence, can be specified based on the linear frame of reference given by H1 with *st.p.(H1)* preceding all locations on the hallway-line (cf. Eschenbach, 1999). This provides an ordering of the rooms, or, in the words of Myers and Konolige (1995), a "sequence of openings," i.e. an ordering of the doors, which is the basis of the *counting-the-relevant-entities-passed strategy*.

The use of LEFT_OF and RIGHT_OF presupposes a planar structure, i.e. a spatial arrangement in which not all points are on one line. If we have a configuration consisting of a line and of points not on the line, we can distinguish configurations of points 'on the same side of the line' and 'on different sides of the line'. Using geometric terms this is expressed by two spatial relations basic in spatial cognition and perception (see Eschenbach and Kulik, 1997, and Eschenbach, 1999). The equivalence relation ON_THE_SAME_SIDE_OF determines two entities as being on the same side of a line. In contrast, ON_DIFFERENT_SIDES_OF specifies that two entities are separated by the line in question.[15]

In general, a line separates the plane in the line itself and two half-planes. To characterize these half-planes as the *right side* or the *left side* we have to distinguish them by labelling one of the sides, i.e. of the equivalence classes. Using a distinguished point as the *right_point* (cf. Figure 15) determines what the right half-plane is.[16] Thus, seeing the hallway H1 as the linear backbone of the planar floor structure starting with *st.p(H1)* and by introducing a label for the right side, we have information enough to interpret the pictorial elements in MAP.3 also with respect to those spatial relations, which require a direction (IN FRONT OF / BEHIND) or a distinguished half plane (LEFT OF / RIGHT OF). For other types of use of directional terms, e.g., the relative one exemplified above with *"left of Barbara's office"*, other lines provide the frame of reference. In the present case, this can be either the 'map user's line of vision', or a line induced by a representational entity that corresponds to the addressee (see Levinson, 1996, ans Eschenbach, 1999, on types of frames of reference).

---

[15]These spatial relations can easily extended to be based on *curves* instead of lines. Curves, axiomatically characterized in Eschenbach et al. (1999), are linear structures, which have not to be *straight* and which can—in contrast to geometrical lines—possess *end-points*. Thus, the analyses presented in this paper for hallways and streets are transferable to rivers, coastal lines, mountain ranges, etc.

[16]The other side is then implicitly labelled as the left side. Which of the two sides is labelled is just a matter of convention.

4   TOWARDS A FORMAL THEORY OF COMMITMENTS IN MAPS

In this final section I will pick up some threads from the prior sections and weave them into a summarizing conception. Let us start from the view that maps can be seen as pictorial representations realized on a planar medium intended as means for communicating information about the spatial environment. A central task in making use of a sketch map for solving a spatial problem is to decide which graphical entities and which configurations of such entities bear content effective and suitable for solving the problem, and, on the other hand, which aspects of the depiction do not represent the environment in an adequate manner.

4.1   The principle of communicative nonaccidentalness

In particular, map users have to detect which parts of the pictorial representations bear content to be conveyed from the map producer by means of the map to the map user (cf. the examples discussed in Section 2). A successful strategy to distinguish what matters is to follow a principle essential for the human visual system, namely, the *Principle of Nonaccidentalness*—also called *Rejection-of-Coincidence Principle* (Rock, 1983)— which Palmer (1999) characterizes as "the hypothesis that the visual system avoids interpreting structural regularities as arising from unlikely accidents of viewing" (p. 299). This strategy has been exemplified with the T-junction drawings discussed in Section 2: whereas a bend in a free-hand sketch has a high probability to be drawn accidentally (cf. Figure 5b), it is not plausible to assume that the kinks in a diagram produced by a graphic tool are coincidental (cf. Figure 5e). In other words, it is suitable to assume that the streets represented in Figure 5e posses kinks as depicted.

In interpreting the Upham maps (Figure 7a and 7b) similar problems of detecting cases of nonaccidentalness appear. I will exemplify this with different types of interrelations between the Freeway depicting line, which is distinguished by its thickness, and other lines representing streets. There are four types of such constellations, namely, (a) such explicitly specified as exits, e.g., Freeway and Mission St., (b) intersections of lines, e.g., Freeway and street between Arrellaga and Sola (Figure 7b), (c) meets/junctions of lines, e.g., Freeway and the vertical, non-labelled street west of De La Vina (in both maps) or Freeway and Santa Barbara in Figure 7b, and, (d) those constellations without common points between street-depicting lines. This variety of constellations suggests to the map user that the production of a specific one is nonaccidental. For example, although State Street allows reaching the area south of Freeway 101, both Upham maps invite the inference that most streets depicted by vertical lines end north of the Freeway and none of these streets crosses Freeway 101. On the other hand, the nonveridically depicted line configurations are outside the area that the map producer seems to see as relevant to get from the Freeway to the Upham. Thus, inviting flawed inferences in other spatial tasks can be seen as minor fault of the map producer.

The original *Principle of Nonaccidentalness* as used in Vision Science (cf. Palmer, 1999) concerns coincidences of visual perception exclusively, e.g. the perspective of the perceiver, position and orientation of the entities perceived and the conditions of lights. In contrast to this, I propose for graphical communication a *Principle of Communicative Nonaccidentalness*: People interpreting maps are allowed to assume that structural irregularities in the map are nonaccidental, i.e. they are intended by the map producer. Thus they bear content and reflect properties of the represented environment. The discussion of the T-junction example demonstrates that the map user's assumptions on the map production process are fundamental in deciding what counts as an irregularity.

4.2 Formal characterization of structural regularities in maps

Geometry is that branch of mathematics that deals with the properties of space and objects in space in general. Thus, maps, which are pictorial representations realized in space, namely, on a planar medium, e.g. paper or a computer screen, are treated in the present section as subject of geometric analysis. The basic methodological conception of the geometric analyses I presented in this article is that of an *axiomatic system*, consisting of a set of concepts and a set of (groups of) axioms characterizing, i.e. constraining, the relationship between the concepts in question (cf. Tarski, 1965, on the axiomatic method in general, and Habel and Eschenbach, 1997, on spatial concepts and their axiomatic characterization; the axiomatic systems used in the present section are described in detail in Eschenbach and Kulik, 1997, and Eschenbach, 1999).

The most basic geometry, the one I start with, is *linear incidence geometry (LIG)*. The basic concepts of *LIG* are given by a set Π of *points*, a set Λ of *lines*, and the relation ι of *incidence*. The set of *incidence axioms*, *L-Inc*, specifies what it means for points to be incident with a line. Using the prefix L points to the range of the axioms, namely that they characterize ordering on linear structures. To sum up in a formal way:

$$LIG =_{def} \langle \{\Pi, \Lambda, \iota\}, \{L\text{-}Inc\}\rangle$$

Expanding *LIG* with additional concepts and axioms leads to other axiomatic systems. By considering the concepts of *ordered line* (Ω) and *precedence* relation ($<$) and adding further *axioms*, on the one hand, *L-Ord*, the set of axioms which specify the preference relation, and on the other hand, *L-Inc-Ord* those axioms which characterise the interaction of incidence and preference, we obtain *linear ordering geometry*:

$$LOG =_{def} \langle \{\Pi, \Lambda, \Omega, \iota, <\}, \{L\text{-}Inc, L\text{-}Ord, L\text{-}Inc\text{-}Ord\}\rangle$$

Pictorial representations are processed using a collection of inspection and manipulation processes, which exploit the graphical elements realized on a spatial medium. For instance, in visual perception and spatial cognition there are processes for the detection of *linearity*, of *incidence* and *precedence*, as well as, of *connection* and *neighbourhood*. I will exemplify this with the Hallway example and the analyses presented in Section 3. During the first steps of processing the map MAP.2—whether the horizontal or the vertical instance—the interpreter brings on the linear structures depicted in Figure 12. Furthermore, the precedence structure of the lines is detected, i.e. MAP.2 is recognized as *possessing LOG structure*, which can be represented by assigning the label *LOG* to the map. Thus, as a *LOG*-map, MAP.2 is suitable for specific spatial concepts, namely, BETWEEN, ADJACENT and OPPOSITE, and the inferences connected with these concepts. For example, with fixed first or last argument in a betweenness constellation, it is allowed to infer from BETWEEN(A, B, C) and BETWEEN(A, C, D) to BETWEEN(A, B, D). Such a constellation is depicted in the grid graph of the hallway (Figure 13), where h12 (area near

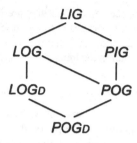

Figure 16: Containment structure of geometries

the doors to R1 and R4) lies between h11 (one end of the hallway) and h13 (area near the doors to R2 and R5), i.e. BETWEEN(h11, h12, h13). Since also BETWEEN(h11, h13, h15) holds, i.e. h13 lies between one end of the hallway, h11, and the other end, h15, it can be inferred, that h12 lies between h11 and h15, and analogously for other constellations.

Furthermore, ordered lines and curves can provide a frame of reference as an anchor for using spatial concept such as IN_FRONT_OF or BEHIND. Specific information in maps can lead to see some lines as directed. For example, in interpreting the structure of MAP.3 (Figures 14), h11 can be seen as the starting point of the hallway—instead of being seen as one end of the hallway—and, thus, the horizontal lines of the grid graph (Figure 13) would be interpreted as lines possessing a *direction* in another geometric structure, LOG_D.

In contrast to interpreting MAP.3 via an ensemble of linear structures—as depicted by the grid graph—MAP.3 can be seen as a structure of a *planar* geometry. The first step to do specifies the range of application: *P-Inc* contains a incidence axiom which requires that for every line points not incident with the line exist, i.e. their range of application is not linear. *Planar incidence geometry* is given by

$$PIG =_{def} \langle \{\Pi, \Lambda, \Omega, \iota\}, \{\textit{L-Inc}, \textit{P-Inc}\} \rangle$$

The combination of the conceptions of *planarity* and *ordering* leads to *planar ordering geometry*, *POG*. Based on the concepts of *LOG* the additional ternary relations *same side (of a line)* and *different sides (of a line)* are defined. The interaction of incidence and ordering is specified by the *Axiom of Separation*, which makes up *P-Inc-Ord*. Thus *planar ordering geometry* is given by

$$POG =_{def} \langle \{\Pi, \Lambda, \Omega, \iota, <\}, \{\textit{L-Inc}, \textit{L-Ord}, \textit{L-Inc-Ord}, \textit{P-Inc}, \textit{P-Inc-Ord}\} \rangle$$

The combination of planar concepts of *POG* and directedness allows us to establish a *right_point* (cf. Figure 15) in POG_D. Thus, spatial concepts as RIGHT_OF and LEFT_OF as well as inferences with respect to these concepts are suitable in POG_D.

Figure 16 shows the system of geometries described above. The inclusions between the sets of axioms and sets of geometric concepts of the axiomatic systems correspond to the top-to-bottom paths through the system. Thus, the *weaker-stronger* relation between geometries is depicted by top-to-bottom lines.

In comprehending a map, interpreters have to identify the relevant geometric structures of the pictorial representation. A suitable comprehension strategy—I call this the *strategy of commitment economy*—is to start with the least committed geometry, which is compatible with the map. Taking a processing perspective, this can be described as follows: interpretation starts with an initial assignment of a geometry to the map in question, for example focusing on ensembles of lines suggests to start at the top of the system (Figure 16), i.e. with *LIG* as a basic geometry for specifying pictorial representations containing points and lines. If there are specific reasons, e.g., prominent pieces of information, the interpretation process changes to another geometry with additional concepts or additional axioms. In the floor plans discussed in Section 3, the prominent linear structure of the hallway and the rooms leads immediately to *LOG*. Furthermore, the depiction of the entrance area of MAP.3 enforces the change to a geometry with the concept of *direction*, namely LOG_D. Additionally, widening the focus from the hallway, which is the dominant linear structure of the floor, to the whole floor as a planar entity, leads to a further shift in the structure of geometries, namely to POG_D. This sequence of geometries assigned to the map—each stronger than the preceding—corresponds to a sequence of stricter commitments: whereas *LOG* is not committed with respect to directional concepts as IN_FRONT_OF or LEFT_OF (and their converse relations), LOG_D and its successors are committed to IN_FRONT_OF and BEHIND, and POG_D additionally to LEFT_OF and RIGHT_OF.

Since the incremental passage through the containment structure of geometries, which is performed in map comprehension, seems regularly to go from less committed, weaker geometries to more committed, stronger geometries, the map producer has to avoid any possibly misleading hint to change to a more committed geometry. For example, highlighting of only one of two staircases in a building, can tempt map users into selecting POG_D to take a distinguished perspective from this staircase.

5 CONCLUSION

In this paper I have described how principles of systematic pragmatics developed by Grice and Levinson can be applied to pictorial communication to solve the notorious problem of indeterminacy of pictorial representations. This problem has high relevance for map design, since during map interpretation the user has to detect the 'grade of determination' of the graphical entities building the map. In other words, the map user has to find out the representational commitments that the map producer coded intentionally.

In Section 3 the geometric structure of pictorial representations, e.g., of maps and diagrams, has been specified using the axiomatic method. Based on this conception, representational commitments in maps can be seen as the assignment of a specific geometry to a map. This assignment establishes in which way—or, with respect to which aspects—the map is determined. Successful communication using maps depends on the users selection of the most appropriate geometry, i.e. that geometry, which allows that inferences, which correspond to the producer's commitments.

This way to solve the indeterminacy problem has been exemplified with respect to concepts of incidence and ordering geometry fundamental for the characterization of directional prepositions as *left–right* and *in front of–behind* is also applicable to other types of determination. For example, most sketch maps used in this paper do only hold rough information about metric properties. Thus, it would not be appropriate to measure—even by approximation—the area of rooms and the hallway, but it could be sensible to use a *rough estimation*, based on an *order of magnitude geometry*.[17]

ACKNOWLEDGEMENTS

The research reported in this paper was partially supported by a grant (project 'Axiomatics of Spatial Concepts', Ha 1237/7) of the Deutsche Forschungsgemeinschaft (DFG) in the priority program on 'Spatial Cognition'. I thank Carola Eschenbach for important comments on prior versions of this paper and for many fruitful discussions on Spatial Cognition in general. Furthermore, the discussions with the participants of the 'Meeting on Fundamental Questions in GIScience' (Manchester, July 2001) were very helpful.

REFERENCES

Allen, G. L. (1997). From knowledge to words to wayfinding: Issues in the production and comprehension of route directions. In Hirtle, S. C. and Frank, A. U. (Eds), *Spatial information theory: A theoretical basis for GIS*, volume 1329 of *Lecture Notes in Computer Science*, pp. 363–372. Berlin: Springer-Verlag.

Barwise, J. and Etchemendy, J. (1995). Heterogeneous logic. In Glasgow, J., Narayanan, H., and Chandrasekaran, B. (Eds), *Diagrammatic Reasoning: Cognitive and Computational Perspectives*, pp. 211–234. Cambridge, MA: MIT Press.

Clark, H. H. (1996). *Using Language*. Cambridge: Cambridge University Press.

Denis, M. (1997). The description of routes: A cognitive approach to the production of spatial discourse. *Cahiers de Psychologie Cognitive*, 16:409–458.

Eschenbach, C. (1999). Geometric structures of frames of reference and natural language semantics. *Spatial Computation and Cognition*, 1:329–348.

Eschenbach, C., Habel, C., and Kulik, L. (1999). Representing simple trajectories as oriented curves. In Kumar, A. N. and Russell, I. (Eds), *FLAIRS-99, Proceedings of the 12th International Florida AI Research Society Conference*, pp. 431–436, Orlando, FL.

Eschenbach, C. and Kulik, L. (1997). An axiomatic approach to the spatial relations underlying left–right and in front of–behind. In Brewka, G., Habel, C., and Nebel, B. (Eds), *KI-97 – Advances in Artificial Intelligence*, Berlin. Springer-Verlag.

Fontaine, S. and Denis, M. (1999). The production of route instructions in underground and urban environments. In Freksa, C. and Mark, D. M. (Eds), *Spatial Information Theory: Cognitive and Computational Foundations of Geographic Science*, volume 1661 of *Lecture Notes in Computer Science*, pp. 83–94. Berlin: Springer-Verlag.

Frank, A. U. (2000). Communication with maps: A formalized model. In Freksa, C., Brauer, W., Habel, C., and Wender, K. F. (Eds), *Spatial Cognition II (Int. Workshop on Maps and Diagrammatical Representations of the Environment, Hamburg)*, volume 1849 of *Lecture Notes in Artificial Intelligence*, pp. 80–99. Berlin: Springer-Verlag.

---

[17] Such *order of magnitude geometry* does—as far as I know—not exist up to now. Therefore, it will be in the focus of our future research.

Glasgow, J. (1993). The imagery debate revisited: A computational perspective. *Computational Intelligence*, 9:309–333.

Glasgow, J., Narayanan, H., and Chandrasekaran, B. (Eds) (1995). *Diagrammatic Reasoning: Cognitive and Computational Perspectives*. Cambridge, MA: MIT Press.

Glasgow, J. and Papadias, D. (1992). Computational imagery. *Cognitive Science*, 16:355–394.

Grice, H. P. (1975). Logic and conversation. In Cole, P. and Morgan, J. C. (Eds), *Speech Acts*, pp. 41–58. New York: Academic Press.

Grice, H. P. (1989). *Studies in the Way of Words*. Cambridge, MA: Harvard University Press.

Gurr, C. (1999). Effective diagrammatic communication: Syntactic, semantic and pragmatic issues. *Journal of Visual Languages and Computing*, 10:317–342.

Habel, C. (1987). Cognitive linguistics: The processing of spatial concepts. *T. A. Informations (Bulletin semestriel de l'ATALA, Association pour le traitement automatique du langage)*, 28:21–56.

Habel, C. (1994). Discreteness, finiteness, and the structure of topological spaces. In Eschenbach, C., Habel, C., and Smith, B. (Eds), *Topological Foundations of Cognitive Science. Papers from the Workshop at the FISI-CS, Buffalo, NY*, pp. 81–90, Hamburg. Graduiertenkolleg Kognitionswissenschaft Hamburg.

Habel, C. and Eschenbach, C. (1997). Abstract structures in spatial cognition. In Freksa, C., Jantzen, M., and Valk, R. (Eds), *Foundations of Computer Science – Potential – Theory – Cognition*, pp. 369–378. Berlin: Springer-Verlag.

Habel, C., Pribbenow, S., and Simmons, G. (1995). Partonomies and depictions: A hybrid approach. In Glasgow, J., Narayanan, H., and Chandrasekaran, B. (Eds), *Diagrammatic Reasoning: Cognitive and Computational Perspectives*, pp. 627–653. Cambridge, MA: MIT Press.

Haugeland, J. (1987). An overview of the frame problem. In Pylyshyn, Z. (Ed), *The Robot's Dilemma: The Frame Problem in Artificial Intelligence*, pp. 77–93. Norwood, NJ: Ablex.

Hilbert, D. (1956). *Grundlagen der Geometrie*. Stuttgart: Teubner, 8th edition. With revisions and supplements by P. Bernays. (Transl.) *Foundations of Geometry*. 2nd edition. La Salle, IL: Open Court. 1971.

Ioerger, T. R. (1994). The manipulation of images to handle indeterminacy in spatial reasoning. *Cognitive Science*, 18:551–593.

Johnson-Laird, P. N. and Byrne, R. M. J. (1991). *Deduction*. Hillsdale, NJ: Lawrence Erlbaum.

Klein, W. (1982). Local deixis in route directions. In Jarvella, R. J. and Klein, W. (Eds), *Speech, Place, and Action*, pp. 161–182. Chichester: Wiley.

Kosslyn, S. M. (1980). *Image and Mind*. Cambridge, MA: Harvard UP.

Kosslyn, S. M. (1989). Understanding charts and graphs. *Applied Cognitive Psychology*, 3:185–226.

Kosslyn, S. M. (1994). *Image and Brain*. Cambridge, MA: MIT Press.

Kulik, L. and Klippel, A. (1999). Reasoning about cardinal directions using grids as qualitative geographic coordinates. In Freksa, C. and Mark, D. M. (Eds), *Spatial Information Theory: Cognitive and Computational Foundations of Geographic Science*, volume 1661 of *Lecture Notes in Computer Science*, pp. 205–220. Berlin: Springer-Verlag.

Landau, B. and Jackendoff, R. (1993). "What" and "where" in spatial language and spatial cognition. *Behavioral and Brain Sciences*, 16:217–238, 255–266.

Levinson, S. C. (1996). Frames of reference and Malyneux's question: Crosslinguistic evidence. In Bloom, P., Peterson, M. A., Nadel, L., and Garrett, M. F. (Eds), *Language and Space*, pp. 109–169. Cambridge, MA: MIT Press.

Levinson, S. C. (2000). *Presumptive Meanings—The Theory of Generalized Conversational Implicature*. Cambridge, MA: MIT Press.

Lovelace, K. L., Hegarty, M., and Montello, D. R. (1999). Elements of good route directions in familiar and unfamiliar environments. In Freksa, C. and Mark, D. M. (Eds), *Spatial Information Theory: Cognitive and Computational Foundations of Geographic Science*, volume 1661 of *Lecture Notes in Computer Science*, pp. 65–82. Berlin: Springer-Verlag.

MacEachren, A. M. (1995). *How Maps Work. Representation, Visualization, and Design*. New York: Guilford Press.

Myers, K. and Konolige, K. (1995). Reasoning with analogical representations. In Glasgow, J., Narayanan, H., and Chandrasekaran, B. (Eds), *Diagrammatic Reasoning: Cognitive and Computational Perspectives*, pp. 273–301. Cambridge, MA: MIT Press.

Oberlander, J. (1996). Grice for graphics: Pragmatic implicature in network diagrams. *Information Design Journal*, 8:163–179.

Palmer, S. E. (1978). Fundamental aspects of cognitive representations. In Rosch, E. and Lloyd, B. (Eds), *Cognition and categorization*, pp. 259–303. Hillsdale, NJ: Erlbaum.

Palmer, S. E. (1999). *Vision Science. Photons to Phenomenology*. Cambridge, MA: MIT Press.

Rock, I. (1983). *The Logic of Perception*. Cambridge, MA: MIT Press.

Simon, H. (1980). *The Science of the Artificial*. Cambridge, MA: MIT Press, 2nd edition.

Slocum, T. A., Blok, C., Jiang, B., Koussoulakou, A., Montello, D. R., Fuhrmann, S., and Hedley, N. R. (2001). Cognitive and usability issues in geovisualization. *Cartography and Geographic Information Science*, 28:61–75.

Tarski, A. (1959). What is elementary geometry? In Henkin, L., Suppes, P., and Tarski, A. (Eds), *The Axiomatic Method, with Special Reference to Geometry and Physics*, pp. 16–29. Amsterdam: North-Holland.

Tarski, A. (1965). *Introduction to Logic and to the Methodology of Deductive Sciences*. New York: Oxford University Press, 3rd edition.

Tschander, L., Schmidtke, H. R., Eschenbach, C., Habel, C., and Kulik, L. (2002). A geometric agent following route instructions. In Freksa, C., Brauer, W., Habel, C., and Wender, K. (Eds), *Spatial Cognition III*. Berlin: Springer-Verlag. In print.

Tversky, B. and Lee, P. U. (1999). On pictorial and verbal tools for conveying routes. In Freksa, C. and Mark, D. M. (Eds), *Spatial Information Theory: Cognitive and Computational Foundations of Geographic Science*, volume 1661 of *Lecture Notes in Computer Science*, pp. 51–64. Berlin: Springer-Verlag.

Tye, M. (1991). *The Imagery Debate*. Cambridge, MA: MIT Press.

Worboys, M. F. (2002). Communicating geographic information in context. In Duckham, M., Goodchild, M. F., and Worboys, M. F. (Eds), *Foundations in Geographic*

Information Science, pp. 33–45. London: Taylor & Francis.

Wunderlich, D. and Reinelt, R. (1982). How to get there from here. In Jarvella, R. J. and Klein, W. (Eds), *Speech, Place, and Action*, pp. 183–201. Chichester: Wiley.

Zhang, J. (1997). The nature of external representations in problem solving. *Cognitive Science*, 21:179–217.

Zhang, J. and Norman, D. A. (1994). Representation in distributed cognitive tasks. *Cognitive Science*, 18:87–122.

CHAPTER 6

Granularity in Change over Time

John G. Stell
School of Computing
University of Leeds, Leeds, LS2 9JT, UK

1 INTRODUCTION

1.1 Evolution over time

We can describe the world, at any one time, as consisting of entities which may possess attributes, such as colour, spatial location, age, and so on. When we consider two times we find relationships between the entities at one time and those at another. For example, a piece of land may be divided into two, the two parts thus being related to the original one; a child is related to its parents; two entities at different times may be regarded as being "the same person"; what was once divided may coalesce into a single entity at a later time. These thoughts lead to a view of entities evolving over time which suggests the picture of Figure 1.

The informal idea of entities evolving or being related to other entities, or being regarded as the same entities at different times, appears natural, and formally this looks like a relation. To examine what properties the relation has, consider the specific example illustrated in Figure 2. Here we have a portion of land at three times. At the first time the land is divided into two separate fields, A and B. By time 2, these two have been amalgamated into a single field, C, which is later subdivided in a different way to give fields

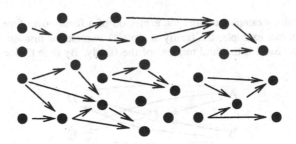

Figure 1: Informal Idea of Evolution

95

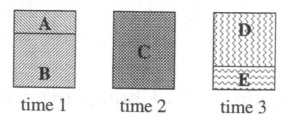

time 1   time 2   time 3

Figure 2: Field Division Example

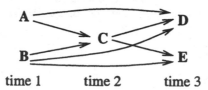

time 1   time 2   time 3

Figure 3: Overlap relation

D and E. Overall there are five entities A, B, C, D, E, and in making a theory of change over time it is necessary to identify what relations exist between the entities at the various times.

One possible relationship between the entities in this spatial example, is to relate x to y if x precedes y and they overlap. This relation is shown diagrammatically in Figure 3. Note that this relation is not transitive: A overlaps C and C overlaps E, but A does not overlap E. There are however other relations between the entities. One possibility is to relate x to y if x precedes y and x is present in the history of y. We can justify the transitivity of the relation in this example. A is involved in the history of E in the sense that the existence of E depends upon the existence of C from which it was created and C was itself created from two entities one of which was A. This relation is shown in Figure 4.

To give another example consider the evolution of a family over time as illustrated in Figure 5. In this example, the family at time t1 consists of just Jim and Sue. At time t2 their son, Sam, is the third member of the family. By time t3 Sue and Jim have

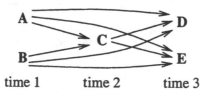

time 1   time 2   time 3

Figure 4: Historical involvement relation

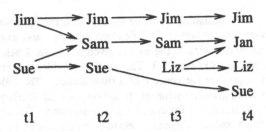

Figure 5: Family example

separated and Sue is not considered part of the family, but Liz, Sam's partner is. By
the final time t4, Sam has died, but Sue has rejoined the family and Jan, the daughter of
Sam and Liz, is also present. Figure 5 provides a convenient diagram, but some of the
relationships are indicated rather than being shown explicitly. For example, Jim at time
t1 is the same person as Jim at time t4, but there is no arrow between these entities in
the diagram. However, the connection can be deduced from the sequence of three three
separate arrows between the entities. This works in this case because the relation of either
being the same person as or being an ancestor of, is transitive.

In the formal development, we use a relation called support, which covers the examples of Figures 4 and 5. The idea being that an entity *a* supports an entity *b* if *a* exists
before *b* and the existence of *a* is necessary for the existence of *b*. The sense in which
'necessary' is meant is not fixed by the theory, but can vary from one application of it to
another. However it is interpreted, the resulting relation will be transitive. If *a* supports
b and *b* supports *c*, then *a* supports *c*, for *c* cannot exist without *a* as *a* is needed for the
existence of *b* and *b* is needed for *c*.

1.2 Overview of the chapter

The topic of this chapter is the development of a formal model of evolving entities and,
in particular, how the model can be given at various levels of detail. The ultimate application of the theory is towards spatio-temporal data in general, and geographical data in
particular, but the earlier parts of the development are independent of the particular kinds
of entities which change over time.

There have been many proposals for the classification of types of change which can
occur in particular kinds of spatio-temporal data. Examples include changes in the content
and presentation of geographic data when it is viewed at various levels of detail, changes
in boundaries of land parcels in cadastral systems, and changes in relationships between
relations between spatial regions when the regions are subjected to continuous change.
The motivation for work of this type is to provide fundamental building blocks out of
which all changes of a particular kind can be constructed. Such fundamental building
blocks can be important both at the conceptual level, for the analysis of processes of
change, and in a practical context, as a means of guiding the implementation of systems
which record changing data. This chapter demonstrates that the concepts of amalgamation
and selection can be used as the basis of a theory of granularity in spatio-temporal data.

A formal model is built up, starting from just abstract sets, adding in variation over time, and adding in further features such as classifications of objects and spatial attributes.

An overview of related work in the area is given in section 2 below. The technical content of the chapter starts in section 3. That section provides the formalization of dynamic sets, which depend on the concept of a time domain. The following section, 4, contains a discussion of granularity for sets based on Stell and Worboys (1999). This is extended to deal with granularity for dynamic sets in section 5. The next two sections, 6 and 7, extend the basic concept of dynamic sets to cover the cases where the entities have a thematic classification, and where they have a spatial location. Some aspects in these two sections are only sketched, and more work will be required to provide a complete picture. The chapter concludes with a short summary and suggestions for further work.

2 RELATED WORK

The subject of cartographic generalization has been widely studied, and has produced schemes which classify types of change in the representation of data on maps at different scales. The focus of this chapter is however, not on the changes to which representations may be subjected, but on changes to the data itself. This is usually referred to as **semantic generalization**.

Timpf in her thesis (Timpf, 1998) identified three kinds of hierarchies: aggregation, filtering and generalization. Each of these types of hierarchy corresponds to a means by which less detailed data may be formed. In aggregation a set of entities is formed into subsets of entities which are indistinguishable at the coarser level of detail. In filtering some entities are selected from a set, and others are omitted. Generalization in Timpf's sense is effectively a kind of aggregation of trees, in which nodes are grouped together in a controlled way. Concepts which correspond exactly to Timpf's aggregation and filtering on sets are discussed by Stell and Worboys (1999). They describe notions of amalgamation and selection not only for sets, but also for graphs. Further work on graphs at different levels of detail and their application to discrete spaces, can be found in Stell's papers (Stell, 1999, 2000). Another approach to classifying kinds of change that can occur when level of detail is varied is due to (Puppo and Dettori, 1995). They deal with simplification mappings between cell complexes, and identify various types of simplification.

Changes in relationships between spatial entities can occur over time as the entities move or evolve. The possible transitions between relationships which can be observed when continuous motion is involved have been classified for several systems of spatial relations, including the RCC5 and RCC8 (Cohn et al., 1997). These classifications lead to the continuity networks for these systems, and subgraphs of these networks correspond to the conceptual neighbourhoods originally introduced by Freksa. This work has been extended to deal with regions which are not crisp but have indeterminate boundaries (Clementini and Felice, 1997; Cohn and Gotts, 1996). One analysis of temporal changes in a cadastral application is due to Spéry et al. (1999). Five elementary changes in the boundaries of land parcels are identified.

Hornsby and Egenhofer (1997) have proposed a change description language capable of modelling changes over time. In later work Hornsby and Egenhofer (1999) addressed views at various levels of detail of objects which are subject to change. This uses a lattice of levels of detail in a similar way to Stell and Worboys (1998) stratified map spaces. Most recently Hornsby and Egenhofer (2002) have considered levels of granularity in the description of moving objects.

Medak (1999) has proposed a formal approach to change over time in spatio-temporal databases. This uses four basic operations affecting object identity: create, destroy, suspend and resume. Algebraic rules for these operations are presented, and an implementation in the functional programming language Haskell is given.

In the present book, the most closely related material is found in Bittner and Smith's *Theory of Granular Partitions*. There are clear connections between granular partitions and the classified dynamic sets in section 6 below. However, the classifications used in the present chapter need not have the form of trees which Bittner and Smith require, and their chapter does not address the partitioning of changing entities, although the theory they present is capable of being extended to treat this case. It would certainly seem worthwhile to investigate whether the concepts for a theory of granularity of change presented in this chapter could be combined with the notion of granular partition to provide a unified account of granularity and change.

3 FORMALIZATION OF EVOLVING ENTITIES

To provide a theory of granularity of change, we have to fix first on a model of change at a fixed level of detail, and before this we need a model of time, over which entities may evolve. A simple model of time is given in 3.1 below. This is followed by the introduction of the notion of dynamic set which is a key concept in this chapter. It turns out that the most appropriate way of defining a dynamic set is to introduce, in 3.3, the notion of a presentation, which corresponds directly to convenient diagrams of entities evolving over time. In the formal account, these diagrams, or their formalizations, are not themselves dynamic sets, only a way of describing them, and there are two equivalent ways of defining exactly what a dynamic set is. These two ways are given in 3.4, and section 3.5 shows how fundamental changes to an evolving entity are related to the formal model.

3.1 Time domains

A **time domain** $(T, \leq)$ consists of a finite set T which carries a partial order, $\leq$. If $t, t' \in T$ where $t < t'$ and there is no t'' such that $t < t'' < t'$, then we say that t and t' are **adjacent** time points. Some examples of time domains are shown in Figure 6. Although the examples discussed later in this chapter are of linear time, the formal framework is not restricted to the linear case. Non-linear time can be used to model alternative accounts of the past or alternative projections of the future.

3.2 Dynamic sets

Given a time domain T, a **dynamic set over** T models the idea of a set of objects which evolve over time. The question of when two entities at different times should be regarded as being the same object is a long-standing and difficult philosophical issue (Gallois,

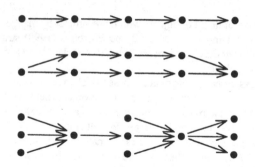

Figure 6: Examples of time domains

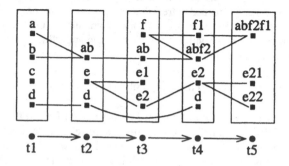

Figure 7: Presentation of a dynamic set

1998). In practical instances we can assume that such issues are settled on the basis of particular application needs and and are not part of a general formal framework such as this. As discussed in the introduction, we do make use of a relationship between entities at different times, which we call the supports relation. The idea is that if entities a and b exist at times t_1 and t_2 respectively and where $t_1 < t_2$, then a supports b if the existence of b depends on the existence on a in some sense. The particular interpretation of the supports relation will depend on the application, it might for example be some kind of causal dependency. There are other possibilities, the coarsest being that any entity supports all later entities. This can be interpreted as a dependency of existence on the view that the state of the world at any time is a consequence of everything that has happened previously.

3.3 Presentation

Consider the picture of Figure 7. The picture shows five successive times t1 to t5, above each of which is a set of entities, to be thought of as the entities existing at that time. The entire set of entities, or more correctly the disjoint union of each of the sets above a time point, bears a relation. Since the relation follows time in moving from left to right in the picture, related pairs are shown joined by a line without an arrow head. Such a picture is a way of presenting the supports relation, and to explain how we need to recall the concept of the covering relation of a poset (Grätzer, 1998, p12).

Given a poset $(A, \leq)$, we define the relation $\prec$ by $a \prec b$ if $a < b$ and there is no x for which $a < x < b$. When A is finite, the relation $\prec$ completely determines the relation $\leq$, the passage from $\prec$ to $\leq$ being given by taking the reflexive transitive closure. The covering relation of a poset will be acyclic, in the sense that in every chain $a_1 \prec a_2 \prec \cdots \prec a_n$ all the a_i are distinct (observe that this includes the property of being irreflexive, i.e. $a \prec a$ never happens). The covering relation is also intransitive, i.e. if $a_1 \prec a_2 \prec a_3$ then $a_1 \not\prec a_3$. Calling a finite relation with these properties a **cover**, then there is a one-to-one correspondence between covers and finite posets, and it is convenient to specify a finite poset by giving its cover.

Thus covers provide a means of presenting posets, and the diagram of Figure 7 consists of two covers, one for the supports relation, and one for the time domain, together with a mapping between them. To describe the mapping we need to consider morphisms between covers. The most general kind of morphism which we require between covers A and B is an order preserving map of relations between A and B^*, the reflexive, transitive closure of B. This is a special case of the morphisms between graphs used in Stell and Worboys (1999), and that reference explains how such morphisms are composed. When dealing with collections of morphisms between structures, the notion of a category is a useful tool, and for basic category theory, the book of Barr and Wells (1995) will be found helpful. The category where the objects are covers and the morphisms are the ones just mentioned, will be denoted **Cover***. We also need to make use of a restricted kind of morphism of covers $f : A \rightarrow B$, where if $a_1 \prec a_2$ in A then $fa_1 \neq fa_2$ in B, these morphisms will be called **irreflexive**.

Now we can define formally a **presentation of a dynamic set** to consist of two covers A and T together with an irreflexive morphism $\rho : A \rightarrow T$. In Figure 7 the effect of the morphism is shown by the vertical alignment in the diagram.

3.4　Formalizations of dynamic set

There are two equivalent ways of formalizing the notion of dynamic set, that is, of specifying what exactly it is that the presentations just discussed present. One, which we introduce first, takes a global set of entities equipped with the supports relation and provides a function assigning to each its time. The second provides for each time a set of entities and instead of one relation, a collection of relations, indexed by pairs of times in temporal sequence.

Dynamic sets have some connections with Kripke models (Mitchell, 1996), but these generally have functions, rather than arbitrary relations between the sets at successive possible worlds (times). The idea of Kripke models where the poset of possible worlds is considered at multiple levels of detail (as in section 5) does not appear to have received much attention in the literature.

3.4.1　Global formulation

A **dynamic set** over a time domain T consists of a poset $(A, \leq)$ and an order-preserving function $\rho : A \rightarrow T$, such that if $\rho a_1 = \rho a_2$ and $a_1 \leq a_2$ then $a_1 = a_2$. The idea behind this restriction is that at a fixed time, an entity supports only itself. Given a presentation $\rho : (A, \prec) \to (T, \prec)$ we obtain a dynamic set $\rho : (A, \leq) \to (T, \leq)$ by taking $(A, \leq)$ and $(T, \leq)$ to be the reflexive-transitive closures of $(A, \prec)$ and $(T, \prec)$ respectively. The function ρ given in the presentation is the same (as a function on the underlying sets) as that required for the dynamic set itself.

3.4.2 Local formalization

A **dynamic set**, X, **over** T consists of for each $t \in T$, a set $X(t)$, and whenever $t \leq t'$, a binary relation $X(t, t')$ between the sets $X(t)$ and $X(t')$. These relations must satisfy the condition that if we have times t, t', t'' with $t \leq t' \leq t''$ and $x \in X(t)$, $x' \in X(t')$, and $x'' \in X(t'')$ with x related to x' by $X(t, t')$ and x' related to x'' by $X(t, t')$, then x must be related to x'' by $X(t, t'')$. We also require that $X(t, t)$ is the identity relation on $X(t)$. These conditions can conveniently be summarized by regarding T as a category, and specifying that X be a lax functor from this category to the category of sets and relations. A useful reference for the concept of lax functor is Kelly and Street (1974).

Given a dynamic set $\rho : A \to T$ in the previous formulation, we can obtain the equivalent local structure by taking $X(t)$ to be $\{a \in A \mid \rho a = t\}$, and defining $X(t, t')$ by restricting the relation on A to the subset of A given by $X(t) \cup X(t')$.

3.5 Fundamental changes

If t and t' are two immediately succeeding times, there are four fundamental changes which can occur between the elements of $X(t)$ and the objects of $X(t')$.

death or suspension An object may be present at t but have no corresponding object at t'. For instance, the element $c \in X(t1)$ in the dynamic set presented in Figure 7, and also $d \in X(t2)$.

birth or resumption An object may be present at t' but have no corresponding object at the previous time. For example $d \in X(t4)$ in Figure 7 and also $e \in X(t2)$.

merge Two or more objects at the earlier stage combine to form an object at the subsequent stage. For example a and b in $X(t1)$ combine to form the object denoted by ab in Figure 7.

split One object at the earlier stage splits into two or more objects at the later stage. This can be seen in the splitting of $e \in X(t2)$ into $e1$ and $e2$ in $X(t3)$.

The absence of each one of these kinds of change corresponds to one of four basic properties which a binary relation may have. If no objects die or suspend, the relation is total; if no objects are born or resume, it is surjective; if no objects merge, it is injective; if no objects split, it is functional.

It is noteworthy that there can be times t, t', t'' where $t < t' < t''$ with $x \in X(t)$ related to some $y \in X(t'')$ but not related to any element of $X(t')$. An instance of this appears with d in Figure 7. This corresponds to having an object which exists at times t and t'' but is in some sense non-existent in between these two times. Well known examples include countries or states which have had multiple episodes of existence through

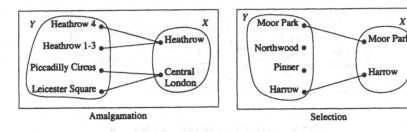

Figure 8: Loss of detail between sets

history, for example Austria. From a formal viewpoint, this behaviour is allowed by permitting X, considered as a functor, to be lax.

4 GRANULARITY FOR STATIC SETS

4.1 Amalgamation and selection

In this section we examine what it means for one set to be a less detailed version of another set. Although abstract sets are the simplest of structures, they provide the basis on which more elaborate models are constructed.

The discussion does not require that the elements of the sets which we consider in this section have any specific properties. However, it is useful for the eventual applications of this work to bear in mind that the elements might be objects which could be stored in a spatial database or a GIS. Examples of such objects could be a specific building, a road, a lake, or a town etc. The data held in a spatial database can be envisaged, at a conceptual level, as a set of objects each of which may be equipped with both spatial and non spatial attributes.

Given a set, X, of objects, a less detailed representation of X will be a set Y formed from X by the combination of two processes.

- Some of the objects of X are selected to contribute to Y, and others are rejected.

- The selected objects may subjected to an equivalence relation so that distinctions between some objects are forgotten.

The two operations are illustrated in Figure 8. In the amalgamation example, the set Y consists of four stations on the London Underground. For some application it may be inappropriate to distinguish between the two individual stations at Heathrow Airport. Similarly, the stations Piccadilly Circus and Leicester Square are physically close together, and at a lower level of detail, the distinction between them may not be important. By avoiding the distinctions between these pairs of stations, we arrive at the set X as a less detailed representation of the data in Y.

The example of selection is also derived from actual data about the London Underground. Here again Y is a set representing four individual stations which is represented at a lower level of detail by a set X containing only two elements. However, in this case

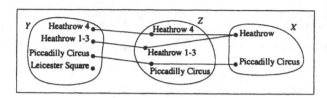

Figure 9: Combined Amalgamation and Selection

the operation performed on Y to produce X is quite different. The stations present in X are selected from those in Y because of their relative importance. Northwood and Pinner are minor stations, and many trains which do stop at Moor Park and Harrow do not stop at the two smaller stations.

When X and Y are sets it is straightforward to formalize the notions of amalgamation and selection. If the relationship of X to Y is one of selection, then there is an injective (or one-to-one) function from X to Y. If the relationship is one of amalgamation, then there is a surjective (or onto) function from Y to X.

4.2   Combining amalgamation and selection for sets

The above examples deal with two ways in which X may be a less detailed representation of Y. In more complicated examples the relationship need not be solely one of amalgamation or selection. In general, a loss of detail relationship between X and Y will involve both selection and amalgamation. This entails a set Z which is obtained from Y by selection, and which is amalgamated to produce X. A simple example appears in Figure 9. A pair consisting of a selection followed by an amalgamation will be called a *simplification* from Y to X. Formally, a simplification from a set Y to a set X consists of three things: a set Z, an injective function from Z to Y (the selection part) and a surjective function from Z to X (the amalgamation part). Alternatively we can describe a simplification from Y to X as a partial surjective function from Y to X.

It might appear that by defining a simplification to consist of a selection followed by an amalgamation, we are being unnecessarily restrictive. It is natural to ask whether this definition of simplification excludes an amalgamation followed by a selection, or a sequence of the form $[s_1, a_1, s_2, a_2, \cdots, s_n, a_n]$ where each a_i is an amalgamation, and each s_i is a selection. In fact, provided we are dealing with simplifications of graphs, or of sets, every sequence of the above form can be expressed as a single selection followed by a single amalgamation. The justification for this lies in the fact that simplifications can be composed.

It is worth noting that a single selection on its own is still a simplification. This is because it can be expressed as a selection followed by the trivial amalgamation in which no distinct entities are amalgamated. Similarly, a single amalgamation on its own is a simplification, since it is equal to the trivial selection, which selects everything, followed by the amalgamation.

4.3 Simplification as amalgamation and selection

A selection from a set X is a set Z and an injective function from Z to X. An amalgamation of a set X is a surjective function from X to a set Z. In general loss of detail can involve both selection and amalgamation, so a **simplification** from X to Y is defined to consist of a set Z, and functions $f : Z \rightarrow X$ and $g : Z \rightarrow Y$ where f is a selection and g is an amalgamation. An alternative description of a simplification from X to Y is a partial function from X to Y which is surjective. This view is often technically simpler, and makes it clear that simplifications can be composed, but it is not as conceptually useful in that twin concepts of selection and amalgamation are not made explicit.

5 GRANULARITY FOR DYNAMIC SETS

The previous section considered simplifications of sets in terms of amalgamation and selection. We now extend this from simple static sets to dynamic ones.

5.1 Simplification of covers

A cover was defined in section 3.3 as a relation on a finite set which is intransitive and acyclic. If $(A, \prec)$ and $(B, \prec)$ are covers, then a function $f : B \rightarrow A$ is a **selection** if it is injective and if $b_1 \prec b_2$ then $fb_1 \prec^+ fb_2$ where $\prec^+$ is the transitive closure of $\prec$. In view of the injectivity condition, we could equivalently put the reflexive-transitive closure in the definition of selection. A function $f : B \rightarrow A$ is an **amalgamation** if it is surjective and $b_1 \prec b_2$ then $fb_1 \prec fb_2$.

A **selection** from a cover $(A, \prec)$ to a cover $(B, \prec)$ is defined to consist of a cover $(C, \prec)$ together with a selection f and an amalgamation g as in the following diagram in the category **Cover***.

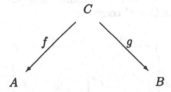

The definition of amalgamation does not allow c_1 to become identified with c_2 if $c_1 \prec c_2$. This is not restrictive when working with simplifications as if an identification is required the selection can be constructed so that the entities to be identified are unrelated by the $\prec$ relation in C.

Describing simplifications in this form is conceptually useful as it emphasizes the separate components of selection and amalgamation. However, it is often useful to use an equivalent description of a simplification from $(A, \prec)$ to $(B, \prec)$ as a partial function from A to B which is surjective and order preserving in the reflexive-transitive closure of the relations, but not necessarily order preserving in the relations themselves.

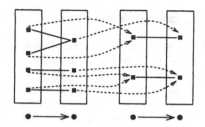

Figure 10: Amalgamation of entities

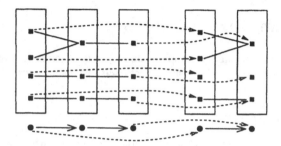

Figure 11: Simplification involving omission of a time point

5.2   Granularity for dynamic sets

We now consider what should be meant by a simplification of a dynamic set. It is easiest
to define this concept in terms of presentations rather than the dynamic sets themselves.

A simplification from (a presentation of) a dynamic set $\rho : A \rightarrow T$ to $\rho' : A' \rightarrow T'$
consists of a dynamic set $\rho'' : A'' \rightarrow T''$, with simplifications of A to A' and T to T' such
that the following diagram in **Cover*** commutes.

$$
\begin{array}{ccccc}
A & \xleftarrow{\;\;\alpha\;\;} & A'' & \xrightarrow{\;\;\sigma\;\;} & A' \\
\downarrow{\scriptstyle\rho} & & \downarrow{\scriptstyle\rho''} & & \downarrow{\scriptstyle\rho'} \\
T & \xleftarrow{\;\;\beta\;\;} & T'' & \xrightarrow{\;\;\tau\;\;} & T'
\end{array}
$$

Some of the situations that are catered for by this definition are illustrated in Figures 10 to 14. In Figure 10 the two time domains are identical. The effect of the simplification on the entities is indicated by the dashed lines. This figure shows that entities of the
same time may be amalgamated by a simplification, and that the images of two entities
unrelated in the support relation may become related after the simplification.

In Figure 11 the simplification of the dynamic sets includes a simplification of the
time domain. In this case only two of the times in the original domain are selected and

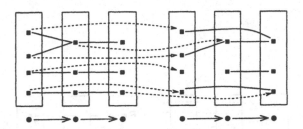

Figure 12: Omission of entities and support connections

the middle one is omitted. In this situation all the entities existing at the omitted time must themselves be omitted too, but the formal definition allows flexibility as to whether relations of support via these omitted entities are maintained in the simplified dynamic set or omitted.

Figure 12 illustrates the omission of entities and of support relations which can occur even when the simplification of the time domains is the identity. Observe in particular that where we have three entities where *a* supports *b* which supports *c*, we may omit *b* and still keep *a* supporting *c*. In terms of the presentations we are working with, this means that the simplified dynamic set presentation will show a link from *a* to *c* even though the original presentation showed no such link. However, in the dynamic sets presented by these presentations *a* supports *c* in both cases.

The formal definition of simplification permits *a* supporting *b* to be simplified to *a* not supporting *b*, but only allows *a* not supporting *b* to be simplified to *a* supporting *b* in very special circumstances, which are illustrated in Figure 13. This asymmetry in which connections of support my be lost, but not in general added fits the notion of simplification. That is, simplification involves the loss of information (in this case that something supports something else) but not the arbitrary addition of new information.

As shown in Figure 13 new connections of support may be created through the requirement that the supports relation is transitive. The Figure presents a situation where neither *a* nor *d* is itself involved in an amalgamation with another entity, and *a* and *d* are initially unrelated in the supports relation. However, since *a* supports *b* and *c* supports *d*, and *b* becomes identified with *c* in the simplification, *a* does support *d* in the dynamic set resulting from the simplification.

Figure 14 shows what can happen when the simplification of the time domain amalgamates two distinct times into one. In this case the definition of simplification of dynamic sets forces some identifications between the entities existing at the times which are identified. If entities *a* and *b* exist at two times which become identified, then *a* and *b* must them selves be identified if we can find a sequence of support connections (ignoring the direction of the supports relation) which link *a* to *b*. The figure also demonstrates that entities taken from each of the sets above the two identified times may or may not become identified themselves when they are not connected by a sequence of support connections.

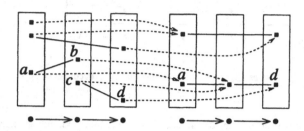

Figure 13: Addition of new support through simplification

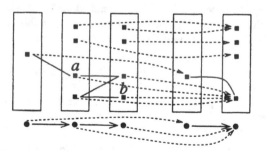

Figure 14: Simplification with amalgamation of two times

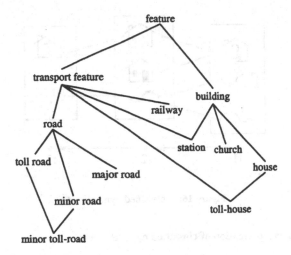

Figure 15: A classification hierarchy

6 CLASSIFIED DYNAMIC SETS

6.1 Classifying sets

So far the objects which vary over a time domain have just been modelled as elements of sets. To introduce more sophisticated models we can add a classification to each object. This allows us to identify some objects as roads, others as railways and yet others as houses. The classification attached to an object can vary over time, for instance a minor road may become a major road, or a railway may become disused and turned into a cycle track. However there are generally limits to changes, for instance a house cannot become a railway, nor a tree a river. The classifications can be arranged into a hierarchy, or possibly several hierarchies, as indicated in Figure 15. A **classification** $(\Phi, \sqsubseteq, \succcurlyeq)$ consists of a set Φ partially ordered by $\sqsubseteq$ and equipped with a reflexive and transitive relation $\succcurlyeq$. It is required that any set of elements of Φ lying in the same connected component with respect to $\sqsubseteq$ has a least upper bound. It is also required that if $\varphi_1 \succcurlyeq \varphi_2$ then there is some ψ such that $\psi \sqsubseteq \varphi_1$ and $\psi \sqsubseteq \varphi_2$. The idea is that if $\varphi_1 \sqsubseteq \varphi_2$ then φ_1 is a more general concept than φ_2, for example animal and dog respectively. The relation $\succcurlyeq$ models the idea that instances of some concept can evolve to instances of another concept. For example, a puppy can over time become a dog but not an elephant, even though all are instances of animal. A **classified dynamic set** is defined to be a dynamic set X over a time domain T together with for each $X(t)$ a function $\eta_t : X(t) \to \Phi$. It is required that if $t \le t'$ in T and $x \in X$ is related by $X(t, t')$ to $x' \in X(t')$ then $\eta_t x \succcurlyeq \eta_{t'} x'$.

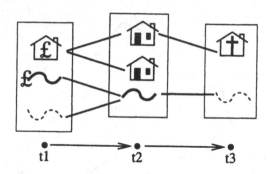

Figure 16: A classified dynamic set

6.2 Temporal simplification of classified dynamic sets

If we have a dynamic set X over a time domain T, and X is classified by Φ, where $\eta_t : X(t) \rightarrow \Phi$, any simplification of time domains $\sigma : T \rightarrow S$ induces a simplification of classified dynamic sets. That is, we can construct a dynamic set Y over S which is also classified by Φ.

To define the dynamic set Y we need to specify

1. For each element s of the time domain S, a set $Y(s)$.

2. For each pair of elements s and s' of S where $s \leq s'$, a relation $Y(s,s')$ between $Y(s)$ and $Y(s')$.

For each $s \in S$, $\sigma^{-1}s$ is a subset of T. Define $W(s)$ to be the disjoint union of the family of sets $(X(t))_{t \in \sigma^{-1}s}$. The set $Y(s)$ is formed from $W(s)$ as the equivalence classes of the equivalence relation which makes $x \in X(t)$ equivalent to $x' \in X(t')$ whenever x and x' are related by the equivalence relation $\sim$, generated by the requirement that $x \sim x'$ if and only if x is related to x' in $X(t,t')$. The relations $Y(s,s')$ are defined by requiring that $y \in Y(s)$ is related to $y' \in Y(s')$ iff there are $x \in y$ and $x' \in y'$ where $x \sim x'$.

All we need to do now is specify how each of these sets is classified. An element of $Y(s)$ is an equivalence class of elements of sets of the form $x \in X(t)$ where $t \in \sigma^{-1}s$. The set $A(s) = \{\eta_t x \in \Phi \mid x \in y,$ and $x \in X(t)\}$ will consist of elements of Φ all lying in the same connected component. Thus the least upper bound of $A(s)$ will exist, and we classify $Y(s)$ by $\theta_s : Y(s) \rightarrow \Phi$ where $\theta_s(y)$ is the least upper bound of $A(s)$.

To illustrate this process, Figure 16 shows an example of a dynamic set classified by the hierarchy in Figure 15. The symbols have the following meaning shown in Figure 17. In the dynamic set at time t1 we have a toll-house, a minor toll-road and a track. By time t2 the toll-house has been divided into two ordinary houses, and the toll-road and track are unified into a single minor road. By time t3, one of the two houses has gone and the other has been converted into a church. The minor road is now merely a track.

Two different induced simplifications of this situation are shown in Figure 18. In the left hand example, times t1 and t2 are amalgamated. At this level of detail, all we can say is that there is a road of some kind and a house. It is not possible to distinguish

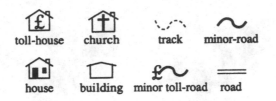

Figure 17: Key to the symbols

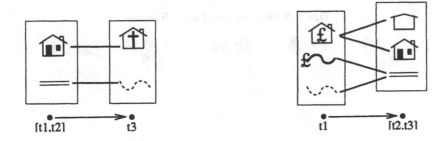

Figure 18: Two induced simplifications

between the three kinds of road which are apparent at the more detailed level. In the right hand example, one of the two houses is amalgamated with the church, and the most detailed classification possible is that there is a building, but the granularity prevents a more detailed classification.

7 SPATIALLY REFERENCED DATA

In a dynamic set X over T, the elements of each $X(t)$ can be given an associated spatial region as well as a classification in a hierarchy. This amounts to adding a temporal dimension to the approach proposed by Stell and Worboys (1998) which was based on the concept of a stratified map space. Thus we use a formal framework where R is some set of regions. These regions might be regular closed sets in the plane, but there are other possibilities, such as the discrete regions discussed in Stell (2000).

In the context of reduction in temporal level of detail egg-yolk regions as discussed in Cohn and Gotts (1996) can be used. An 'egg-yolk' region consists of a pair of regions (x, y) where x is a subregion of y. The original formulation of egg-yolk regions required that x be a proper part of y, but for the application here, it is more appropriate to allow x to be equal to y. The set of egg-yolk regions constructed from a set of regions, R, will be denoted $R^{\circledS}$.

Egg-yolk regions are often used to represent regions with indeterminate boundaries. With this interpretation, (x, y) can be used to stand for a region which we know contains x and which is contained in y, but the precise location of which is uncertain. Egg-yolk

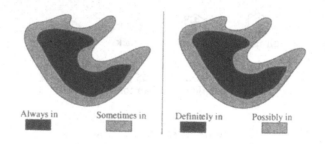

Figure 19: Interpretations of egg-yolk regions

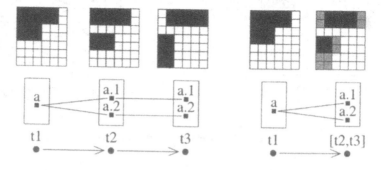

Figure 20: Indeterminacy from spatio-temporal granularity

regions can also be used to represent regions which vary over time. In this interpretation, (x, y) can stand for a region which always contains x and where anywhere in the difference $y - x$ is sometimes but not always in the region. These two interpretations of an egg-yolk region are illustrated in Figure 19. A simple example of how temporally indeterminate regions arise from change in level of detail appears in Figure 20. Here we have a dynamic set X over a three element time domain $T = \{t1, t2, t3\}$. The set $X(t1)$ consists of a single element, a, which has an associated region shown above it in the diagram. By time $t2$, the entity a has divided into a.1 and a.2 each of which has an associated region. By time $t3$ the regions associated to the two entities have moved to a different location. On the right of Figure 20 is shown the result of a simplification of the time domain which amalgamates $t2$ and $t3$. Here the region associated to a.1 is an egg-yolk region. The yolk is the intersection of the regions associated to a.1 at the two times, and the egg is the union of the two regions. The egg-yolk region associated to a.2 is constructed similarly.

To describe the general situation, suppose we have a dynamic set X over a time domain T which is simplified to S. A given $s \in S$ will result from the amalgamation of a set of elements of T, say $Q(s)$. The construction of the simplified dynamic set Y over S from X has been described earlier. The additional structure here is that each $X(t)$ is equipped with a function $\rho_t : X(t) \to R^{\circledcirc}$. We are deliberately using $R^{\circledcirc}$ rather than R here in order to cope with repeated simplifications. Also the decision to allow egg-yolk regions (x, y) where $x = y$ means that any region in R can be regarded as element of $R^{\circledcirc}$, thus the example given above is included in our framework. It remains to show how each $Y(s)$ is equipped with a function $\sigma_s : Y(s) \rightarrow R^{\circledcirc}$. Each element $y \in Y(s)$ will result from the amalgamation of a set of elements each of which belongs to some $X(t)$ where $t \in Q(s)$. For each $t \in Q(s)$ let $A(t, s)$ denote the subset of $X(t)$ which is amalgamated to form y. Then the function σ_s is defined by

$$\sigma_s(y) = \left(\bigcap_{t \in Q(s)} \bigcup_{x \in A(t,s)} \downarrow \rho_t(x), \; \bigcup_{t \in Q(s)} \bigcup_{x \in A(t,s)} \uparrow \rho_t(x) \right)$$

where $\downarrow \rho_t(x)$ and $\uparrow \rho_t(x)$ denote respectively the yolk and the egg of $\rho_t(x)$.

8 CONCLUSIONS AND FURTHER WORK

This chapter has presented a formal model for data varying over time based on a set of entities at each time and relations between these sets. The simple concept of a relation between sets is able to deal with most of the important changes to which entities can be subjected. These include birth, death, suspension of existence, resumption of existence, merging and splitting. By equipping these sets of entities with functions to classification hierarchies and to sets of egg-yolk regions it is possible to deal with entities having spatial attributes and non-spatial classifications.

There are several areas for further work. One topic which could be investigated in this framework would be topological relations between regions varying over time and their description at varying levels of detail. For example, if regions are disjoint at time t1 and partially overlapping at t2 we could ask for a description of their relation when the times t1 and t2 are amalgamated.

Another topic would be the incorporation of spatial indeterminacy into the framework. That is, the regions could not only vary over time, but also be of uncertain location. This would appear to need a more elaborate kind of region than the egg-yolks used in section 7 above.

ACKNOWLEDGMENTS

Much of the work reported here was carried out while the author was at Department of Computer Science, Keele University, and an earlier report on the topic was presented at the ECAI 2000 spatio-temporal workshop in Berlin. The work was supported by an EPSRC grant 'Managing Vagueness Uncertainty and Granularity in Spatial Information Systems'.

REFERENCES

Barr, M. and Wells, C. (1995). *Category Theory for Computing Science*. Prentice Hall, second edition.

Bittner, T. and Smith, B. (2002). A theory of granular partitions. In Duckham, M., Goodchild, M. F., and Worboys, M. F. (Eds), *Foundations in Geographic Information Science*, pp. 117–149. London: Taylor & Francis.

Clementini, E. and Felice, P. D. (1997). Approximate topological relations. *International Journal of Approximate Reasoning*, 16:173–204.

Cohn, A. G., Bennett, B., Gooday, J., and Gotts, N. M. (1997). Qualitative spatial representation and reasoning with the region connection calculus. *GeoInformatica*, 1:275–316.

Cohn, A. G. and Gotts, N. M. (1996). The 'egg-yolk' representation of regions with indeterminate boundaries. In Burrough, P. A. and Frank, A. U. (Eds), *Geographical Objects with Indeterminate Boundaries*, volume 2 of *GISDATA Series*, pp. 171–187. Taylor and Francis.

Gallois, A. (1998). *Occasions of Identity*. Clarendon Press, Oxford.

Grätzer, G. (1998). *General Lattice Theory*. Birkhäuser.

Hornsby, K. and Egenhofer, M. (1997). Qualitative representation of change. In Hirtle, S. C. and Frank, A. U. (Eds), *Spatial Information Theory, International Conference COSIT'97, Proceedings*, volume 1329 of *Lecture Notes in Computer Science*, pp. 15–33. Springer-Verlag.

Hornsby, K. and Egenhofer, M. (1999). Shifts in detail through temporal zooming. In Camelli, A., Tjoa, A. M., and Wagner, R. R. (Eds), *Tenth International Workshop on Database and Expert Systems Applications. DEXA99*, pp. 487–491. IEEE Computer Society.

Hornsby, K. and Egenhofer, M. (2002). Modeling moving objects over multiple granularities. *Annals of Mathematics and Artificial Intelligence*, 36:177–194.

Kelly, G. M. and Street, R. (1974). Review of the elements of 2-categories. In Kelly, G. M. (Ed), *Proceedings of the Sydney Category Theory Seminar 1972/1973*, volume 420 of *Lecture Notes in Mathematics*, pp. 75–103. Springer-Verlag.

Medak, D. (1999). Lifestyles—An algebraic approach to change in identity. In Böhlen, M. H., Jensen, C. S., and Scholl, M. O. (Eds), *Spatio-Temporal Database Management. International Workshop STDBM'99. Proceedings*, volume 1678 of *Lecture Notes in Computer Science*, pp. 19–38. Springer-Verlag.

Mitchell, J. C. (1996). *Foundations for Programming Languages*. MIT Press.

Puppo, E. and Dettori, G. (1995). Towards a formal model for multiresolution spatial maps. In *Advances in Spatial Databases SSD'95*, volume 951 of *Lecture Notes in Computer Science*, pp. 152–169. Springer-Verlag.

Spéry, L., Claramunt, C., and Libourel, T. (1999). A lineage metadata model for the temporal management of a cadastre application. In Camelli, A., Tjoa, A. M., and Wagner, R. R. (Eds), *Tenth International Workshop on Database and Expert Systems Applications. DEXA99*, pp. 466–474. IEEE Computer Society.

Stell, J. G. (1999). Granulation for graphs. In Freksa, C. and Mark, D. (Eds), *Spatial Information Theory. Cognitive and Computational Foundations of Geographic Information Science. International Conference COSIT'99*, volume 1661 of *Lecture Notes in Computer Science*, pp. 417–432. Springer-Verlag.

Stell, J. G. (2000). The representation of discrete multi-resolution spatial knowledge. In Cohn, A. G., Giunchiglia, F., and Selman, B. (Eds), *Principles of Knowledge Repre-*

sentation and Reasoning: Proceedings of KR2000, pp. 38–49. Morgan Kaufmann.

Stell, J. G. and Worboys, M. F. (1998). Stratified map spaces: A formal basis for multi-resolution spatial databases. In Poiker, T. K. and Chrisman, N. (Eds), *SDH'98 Proceedings 8th International Symposium on Spatial Data Handling*, pp. 180–189. International Geographical Union.

Stell, J. G. and Worboys, M. F. (1999). Generalizing graphs using amalgamation and selection. In Güting, R. H., Papadias, D., and Lochovosky, F. (Eds), *Advances in Spatial Databases. 6th International Symposium, SSD'99*, volume 1651 of *Lecture Notes in Computer Science*, pp. 19–32. Springer-Verlag.

Timpf, S. (1998). *Hierachical Structures in Map Series*. PhD thesis, Department of GeoInformation, Technical University, Vienna.

CHAPTER 7

A Theory of Granular Partitions

Thomas Bittner and Barry Smith
Institute for Formal Ontology and Medical Information Science
University of Leipzig, Leipzig, 04107, Germany

1 INTRODUCTION

Imagine that you are standing on a bridge above a highway checking off the makes and models of the cars that are passing underneath. Or that you are a postal clerk dividing envelopes into bundles; or a laboratory technician sorting samples of bacteria into species and subspecies. Or imagine that you are making a list of the fossils in your museum, or of the guests in your hotel on a certain night. In each of these cases you are employing a certain grid of labeled cells, and you are recognizing certain objects as being located in those cells. Such a grid of labeled cells is an example of what we shall call a *granular partition*. We shall argue that granular partitions are involved in all naming, listing, sorting, counting, cataloguing and mapping activities. Division into units, counting and parceling out, mapping, listing, sorting, pigeonholing, cataloguing are activities performed by human beings in their traffic with the world. Partitions are the cognitive devices designed and built by human beings to fulfill these various listing, mapping and classifying purposes.

In almost all current work in areas such as common-sense reasoning and natural language semantics it is the naïve portion of set theory that is used as basic framework. The theory of granular partitions as it is developed in this paper is intended to serve as an alternative to set theory both as a tool of formal ontology and as a framework for the representation of human cognition. Kinds, sorts, species and genera are standardly treated as sets of their instances; subkinds as subsets of these sets. Set theory nicely does justice to the granularity that is involved in our sorting and classification of reality by giving us a means of treating objects as *elements* of sets, i.e. as single whole units within which further parts are not recognized. But set theory also has its problems, not the least of which is that it supports no distinction between natural totalities (such as the species *cat*) and such *ad hoc* totalities as, for example, {the moon, Napoleon, justice}.

Set theory has problems, too, when it comes to dealing with time, and with the fact that biological species and similar entities may remain the same even when there is a turnover in their instances. For sets are identical if and only if they have the same members. If we model the species *cat* as the set of its instances, then this means that cats form a different species every time a cat is born or dies. If, similarly, we identify an organism as the set of its cells, then this means that it becomes a different organism whenever cells are gained or lost.

Set theory has problems also when it comes to dealing with the relations between granularities. An organism is a totality of cells, but it is also a totality of molecules, and it is also a totality of atoms. Yet the corresponding *sets* are distinct, since they have distinct members.

More recently, attempts have been made to solve some of these problems by using mereology or the theory of part and whole relations (Smith, 1998) as a framework for ontological theorizing. Mereology is better able to do justice in realistic fashion to the relations between wholes and their constituent parts at distinct levels of granularity. All the above-mentioned totalities (of cells, molecules, atoms) are, when treated mereologically, one and the same. Mereology also has the advantage over set theory when it comes to serving as a tool for the sort of middle-level ontological theorizing which the study of common-sense reasoning requires. For mereology does not require that, in order to quantify over wholes of given sorts, one must first of all specify some level of ultimate parts (the *Urelemente* of set theory) from out of which all higher-level entities are then constructed.

But mereology, too, has its problems. Thus it, too, has no way it has no way of dealing with entities which gain and lose parts over time and it has no way of distinguishing intrinsically unified wholes from *ad hoc* aggregations. Above all, its machinery for coping with the phenomenon of granularity brings problems of its own, for if we quantify over wholes, in a mereological framework, then we thereby quantify over all the parts of such wholes, at all levels of granularity. The selectivity of intentionality means however that we are often directed cognitively to coarse-grained wholes whose finer-grained parts are traced over: when I think of Mary I do not think of all the molecules in Mary's arm. Mereology has no means of mimicking the advantages of set-theory when it comes to dealing with such phenomena, and it is no small part of our project here to rectify this defect. The theory of granular partitions is the product of an effort to build a more realistic, and also a more general and flexible, framework embodying the strengths of both set theory and mereology while at the same time avoiding their respective weaknesses.

2   TYPES OF GRANULAR PARTITIONS

Some types of granular partitions are flat: they amount to nothing more than a mere list. Others are hierarchical: they consist of cells and subcells, the latter being nested within the former. Some partitions are built in order to reflect independently existing divisions on the side of objects in the world (the subdivision of hadrons into baryons and mesons, the subdivision of quarks into *up, down, top, bottom, charm, strange*). Other partitions—for example the partitions created by nightclub doormen or electoral redistricting commissions—are themselves such as to create the corresponding divisions on the side of their objects, and sometimes they create those very objects themselves. Quite different sorts of partitions—having cells of different resolutions and effecting unifyings and slicings and reapportionings of different types—can be applied simultaneously to the same domain of objects. The people in your building can be divided according to gender, social class or social security number. Or they can be divided according to tax bracket, blood type, current location or Erdös number. Maps, too, can impose subdivisions of different types upon the same domain of spatial reality, and the icons which they employ represent objects in granular fashion (which means that they do not represent the corresponding object parts). Maps will turn out to be important examples of granular partitions in the sense intended here.

The theory of partitions is, as will by now be clear, highly general, and this generality brings with it a correspondingly highly general reading of the term 'object'. Here we take an object to be any portion of reality: an individual, a part of an individual, a class of individuals (for example a biological species), a spatial region, a political unit (county, polling district, nation), or even (for present purposes) the universe as a whole. An object in the partition-theoretic sense is everything (existent) that can be recognized by some cell of a partition.

Objects can be either of the bona fide or of the fiat sort (Smith, 2001a). Bona fide objects, for example the moon, your armchair, this piece of cheese, are objects which exist (and are demarcated from their surroundings) independently of human partitioning activity. Fiat objects are objects which exist (and are demarcated from their surroundings) only because of such partitioning activity. Examples are: census tracts, your right arm, the Western Hemisphere.

In some cases partition cells recognize pre-existing fiat objects, in other cases the latter are created through the very projection of partition cells onto a corresponding portion of reality. Examples are the partitions creating the States of Wyoming and Montana, or the partitions of a population into persons belonging to distinct tax brackets created by tax legislation. Once fiat objects have been created in this way subsequent partitions may simply recognize them (without any object-creating effect), just as there are partitions which simply recognize bona fide objects.

Our notion of granular partition is only distantly related to the more familiar notion of a partition defined in terms of equivalence classes. Our partitions can include more structure in the form of hierarchically arranged subcells and supercells. Moreover, it is possible to define a partition in terms of an equivalence relation only where the relevant domain has already been divided up into units (the elements of the set with which we begin). The very process of division into units—for example through the imposition of fiat subdivisions or fiat discretizations upon continuous gradations—is however one of the things which our present theory is designed to illuminate.

In Smith and Brogaard (2002b) the notion of granular partition was introduced as a generalization of David Lewis's (1991) conception of classes as the mereological sums of their constituent singletons. Granular partitions, too, can in first approximation be conceived as the mereological sums of their constituent *cells*. The cells within a granular partition may however manifest a range of properties which the singletons of set theory lack. This is because, where a singleton is defined in the obvious way in terms of its member, each cell of a granular partition is defined by its *label*, and this means: independently of any object which might fall within it. The cells of a partition are what they are independently of whether there are objects located within them. A map of Middle Earth is different from a map of the Kingdom of Zenda, even though there is in both cases precisely nothing on the side of reality upon which these maps would be projected. 'The Morning Star' and 'The Evening Star' were for a long time used as labels for two distinct cells in astronomers' partitions of the heavenly bodies, even though, as it later turned out, it is one the same object that is located in each.

If one thinks that there are dodos, then one makes a different sort of error from the error which one makes if one thinks that there is an intra-Mercurial planet. (Set theory, it almost goes without saying, lacks the machinery to deal with such different sorts of error.)

Just as when we point our telescope in a certain direction we may fail to find what we are looking for, so when we point our partition in a certain direction it may be that there are no objects located in its cells. There may, in this sense, be empty cells within a partition (and even, in the most general version of our theory, partitions all of whose cells are empty). But this does not mean that the theory of partitions recognizes some counterpart of the set theorist's empty set (an entity that is contained as a subset within every set). For the empty set is empty by necessity; a cell in a partition, in contrast, is at best empty *per accidens*; it is empty because of some failure on our part in our attempts to partition the reality beyond.

The theory of partitions is thus more powerful than set theory, in that it is better able to do justice to the various ways in which human beings are related, cognitively, to objects in reality. In many ways, however, partition theory is also much weaker than set theory. For the axioms of set theory imply the existence of an entire hierarchy of sets, sets of sets, and so on, *ad infinitum*, reflecting the fact that they were designed to yield an instrument of considerable mathematical power. Partition theory, in contrast, is like mereology in that it is attuned to purposes other than those of mathematics. More specifically, it is designed to do justice to the sometimes *ad hoc* ways in which cognitive classificatory instruments are constructed by human beings for specific human purposes.

Partition theory differs from set theory also in this: that it puts partitions and objects in two entirely separate realms. Partitions themselves are never objects, and there are no partitions of partitions. Thus partition theory has no counterpart of sets of sets or of the distinction between two ways in which one set can be contained within another (on the one hand as element, on the other hand as subset). Partition theory can thus provide a framework for theorizing about the relations between cognitive artifacts such as lists and maps and the reality to which such artifacts relate in such a way that debates for example concerning the status of the hierarchy of transfinite sets can be avoided.

3   GRANULAR PARTITIONS AS SYSTEM OF CELLS

3.1   A bipartite theory

In the present paper we present the basic formal theory of granular partitions, leaving for a later work the presentation of the theory of cell-labeling (and, more generally, of the cognitive aspects of partitions as we here understand them). Our formal theory has two orthogonal and independent parts: (A) a theory of the relations between cells, sub-cells, and the partitions in which they are contained; (B) a theory of the relations between partitions and objects in reality. The counterpart of (A) in a set-theoretic context would be the study of the relations among subsets of a single set; the counterpart of (B) would be the study of the relations between sets and their members. These set-theoretical counterparts of (A) and (B) are, be it noted, not independent. This is because the standard subset relation of set theory is itself defined in terms of the set-membership relation (x is a subset of y means: all the members of x are members of y). In the context of partition theory, in contrast, the corresponding relations are defined independently of each other. Partition theory thus departs from the extensionalism of set theory (i.e. from the assumption that each set is defined exclusively by its members). A cell is defined by its position within a partition and by its relations to other cells, and it is this which gives rise to the relations treated of by theory (A). What objects in reality are located in a cell—the matter of theory (B)—is then a further question, which is answered, in different ways from case to case. Briefly, we can think of cells as being projected onto objects in something like the way in which flashlights are projected upon the objects which fall within their purview.

Consider the left part of Figure 1. Theory A governs the way we organize cells into nesting structures and the way we label cells. Theory B governs the way these cell-structures project onto reality indicated by the arrows connecting the left and the right parts of the Figure. Our strategy in what follows will be first of all to define a series of master conditions belonging to theory (A) and theory (B) respectively, and which—for the purposes of the present paper—all partitions will be assumed to satisfy. In later sections we will add further conditions, satisfied by some partitions but not by others.

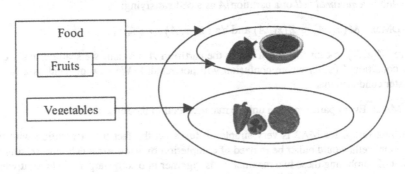

Figure 1: Relationships between cells and objects

3.2 The subcell relation

Theory (A) is effectively a theory of well-formedness for partitions; it studies properties partitions have in virtue of the relations between and the operations performed upon the cells from out of which they are built, independently of any linkage to reality beyond. Cells in partitions may be nested one inside another in the way in which, for example, the species *crow* is nested inside the species *bird*, which in turn is nested inside the genus *vertebrate* in standard biological taxonomies. When one cell is nested inside another in this way we say that the former is a sub-cell of the latter. Note that the subcell relation can hold between two cells independently of whether there are any objects located in them (as for example in relation to the cells labeled 'male dodos' and 'dodos' in a classification of extinct animals).

We use $z, z_1, z_2, \ldots$ as variables ranging over cells and $A, A_1, A_2, \ldots$ as variables ranging over partitions ('cell' is *'Zelle'*, partition is *'Aufteilung'* in German). We write $z_1 \subseteq_A z_2$ in order to express the fact that z_1 stands in a sub-cell relation to z_2 within the partition A. (Where confusion will not result we will drop the explicit reference to the partition A and write simply '$\subseteq$'). We can then state the first of several master conditions on all partitions as follows:

MA1: The subcell relation $\subseteq$ is reflexive, antisymmetric, and transitive.

This means that within every partition: each cell is a subcell of itself; if two cells are subcells of each other then they are identical; and if cell z_1 is a sub-cell of z_2 and z_2 a sub-cell of z_3, then z_1 is in its turn a sub-cell of z_3. We can think of the sub-cells of a cell within a given partition as special sorts of *parts* of the cell; they are those parts which are included within this same partition as cells in their own right.

3.3 Existence of a maximal cell

We define a *maximal cell* of a partition A as a cell satisfying:

DMax: $M(z_1, A) \equiv Z(z_1, A)$ and $\forall z : Z(z, A) \rightarrow z \subseteq z_1$.

Here '$Z(z, A)$' means that z is a cell in the partition A. (Again: we shall normally omit the condition $Z(z, A)$ where confusion will not result.) We now demand as a further master condition that

MA2: Every partition has a unique maximal cell in the sense of DMax.

The motivation for MA2 is very simple: it turns on the fact that a partition with two maximal cells would either be in need of completion by some extra cell representing the result of combining these two maximal cells together into some larger whole; or it would not be one partition at all, but rather two separate partitions, each of which would need to be treated in its own right within the framework of our theory.

We also call the unique maximal cell of a partition its root, $r(A)$. The maximal cell of a partition is such that all the cells in the partition are included in it as subcells. MA2 implies that there are no partitions which are empty *tout court* in that they have no cells at all.

3.4 Finite chain condition

The transitivity of $\subseteq$ generates a nestedness of cells inside a partition in the form of chains of cells satisfying $z_1 \supset z_2 \supset \ldots \supset z_n$, with z_1 as root. We shall call the cells at the ends of such chains *minimal cells* or *leaves*, and define:

DMin: $Min(z_1, A) \equiv Z(z_1, A)$ and $\forall z : Z(z, A) \rightarrow (z \subseteq z_1 \rightarrow z = z_1)$

Another important aspect of a partition is then:

MA3: Each cell in a partition is connected to the root via a finite chain of immediate succeeding cells.

A cell z_2 is the immediate successor of the cell z_1 if and only if $z_1 \subseteq z_2$ and there does not exist a cell z_3 such that $z_1 \subset z_3 \subset z_2$ holds.

MA3 does not rule out the possibility that a given cell within a partition might have infinitely many immediate subcells (also called daughter cells). Enforcing finite chains thus leaves open the issue as to whether partitions themselves are finite.

If, in counting off the cars passing beneath you on the highway, your checklist includes one cell labeled *red cars* and another cell labeled *Chevrolets*, we will rightly feel that there is something amiss with your partition. One problem is that you will almost certainly be guilty of double counting. Another problem is that there is no natural relationship between these two cells, which seem rather to belong to distinct partitions. As a step towards rectifying such problems we shall insist that all partitions must satisfy a condition according to which every pair of distinct cells within a partition stand to each other either in the subcell relation or in the relation of disjointness. In other words:

MA4: If two cells within a partition overlap, then one is a subcell of the other.

Or in symbols:

$$\exists z : (z \subseteq z_1 \text{ and } z \subseteq z_2) \rightarrow z_1 \subseteq z_2 \text{ or } z_1 \supset z_2.$$

(Here and in what follows initial universal quantifiers are taken as understood.) From MA3 and MA4 we can prove by a simple *reductio* that the chain connecting each cell of a partition to the root is unique.

3.5 Partition-theoretic sum and product of cells

The background to all our remarks in this paper is mereology. We take the relation $\leq$ meaning 'part of' as primitive, and define the relation of overlap between two entities simply as the sharing of some common part. $\leq$ is like $\subseteq$ in being reflexive, anti-symmetric and transitive, but the two differ in the fact that $\subseteq$ is a very special case of $\leq$.

The subcells of a cell are also parts of the cell (just as, for David Lewis, 1991, each singleton is a part of all the sets in which it is included). What happens when we take the mereological products and sums of cells existing within a partition? In regard to the mereological product, $z_1 * z_2$, of two cells matters are rather simple. This product exists only when the cells overlap mereologically, i.e. only when they have at least one subcell in common. This means that the mereological product or intersection of two cells, if it exists, is in every case just the smaller of the two cells.

In regard to the mereological *sum* of cells $z_1 + z_2$, in contrast, it is a more difficult situation which confronts us. Given any pair of cells within a given partition, the corresponding mereological sum does indeed exist—simply in virtue of the fact that the axioms of mereology allow unrestricted sum-formation. (This is a trivial matter, for the mereologist: if you got the parts, whatever they are, then you got the whole.) But only in special cases will this mereological sum be itself a cell within the partition in question. This occurs for example when cells labeled 'male rabbit' and 'female rabbit' within a partition have as their sum the cell labeled 'rabbit'. There is, in contrast, no cell in our standard biological partition of the animal kingdom labeled *rabbits and jellyfish*, and there is no cell in our standard geopolitical partition of the surface of the globe labeled *Hong Kong and Algeria*.

To make sense of these matters we need to distinguish the mereological sum of two cells from what we might call their partition-theoretic sum. We can define the former as just the result of taking the two cells together in our thoughts and treating the result as a whole. We can define the latter as follows. The partition-theoretic sum $z_1 \cup z_2$ of two cells in a partition is the smallest subcell within the partition containing both, z_1 and z_2, as subcells; i.e., it is the least upper bound of z_1 and z_2 with respect to $\subseteq$. (By MA2 and MA4 we know that this is always defined and that it is unique.) This partition-theoretic sum is in general distinct from the mereological sum of the corresponding cells. (The partition-theoretic sum of the cells labeled *rabbit* and *lion* is the cell labeled *mammal* in our partition of the animal kingdom.) The best we can say in general is that $z_1 + z_2$ is at least part of $z_1 \cup z_2$ (Smith, 1991). Note, on the other hand, that if we analogously define the partition-theoretic product, $z = z_1 \cap z_2$, of two cells within a given partition as the largest subcell shared in common by z_1 and z_2, i.e., as their greatest lower bound with respect to $\subseteq$, then it turns out that this coincides with the mereological product already defined above.

Mereological sum and product apply to both cells and objects; partition-theoretic sum applies only to cells. Here we use the symbols for the two groups of relations as shown in Table 1:

	Partition-theoretic (for cells)	Mereological (for cells and for objects)
Sum	$\cup$	$+$
Product	$\cap$	$*$
Inclusion	$\subseteq$	$\leq$
Proper Inclusion	$\subset$	$<$

Table 1: Partition-theoretic and mereological relations and operations

When restricted to cells within a given partition $\subseteq$ and $\leq$ coincide, and so also do $\cap$ and $*$. We can think of $\subseteq$ as the result of restricting $\leq$ to the *natural units* picked out by the partition in question. We can think of set theory as amounting to the abandonment of the idea that there is a distinction between natural units and arbitrary unions. Set theory, indeed, derives all its power from this abandonment.

3.6 Trees

Philosophers since Aristotle have recognized that the results of our sorting and classifying activities can be represented as those sorts of branching structures which mathematicians nowadays called trees. Trees are directed graphs without cycles. They consist of nodes or vertices and of directed edges that connect the nodes. That the edges are directed means that the vertices connected by an edge are related to each other in a way that is analogous to an ordered pair. Here we are interested specifically in rooted trees, which is to say: trees with a single topmost node to which all other vertices are connected, either directly or indirectly, via edges. In a rooted tree, every pair of vertices is connected by one and only one chain (or sequence of edges). We shall think of the directedness of an edge as proceeding down the tree from top to bottom (from ancestors to descendants). That a tree is without cycles means that, if we move along its edges, then we will always move down the tree and in such a way that, however far we travel, we will never return to the point from which we started.

The connection between partitions and trees will now be obvious: it is a simple matter to show that every finite partition can be represented as a rooted tree of finite depths and vice versa (Mark, 1978). To construct a tree from a finite partition we create a graph by mapping the cells z_i of the partition onto nodes v_i within the graph and by introducing a directed edge from vertex v_i to v_j if and only if the cell z_i has cell z_j as an immediate subcell. That this is always possible follows from the fact that the subcell relation is well defined (by MA1) and from the fact that chains of immediate cells are always finite (MA3). We can easily show also that the resulting graph is a rooted tree, which follows from MA2; that the graph structure is connected (from MA2), and acyclical (from MA4); and that there is a unique path between any two vertexes (from MA2, MA3 and MA4). The complementary reconstruction of a partition from its tree representation is no less trivial.

We can represent a partition not only as a tree but also as a simple sort of Venn diagram. In a Venn diagram partition cells are represented as topologically simple and regular regions of the plane. Our partitions are Venn diagrams within which regions do not intersect. (Conversely every array of non-intersecting, possibly nested regions in the plain can be transformed into a tree in such a way that each region is represented by a node in the tree, and each directed link in the tree represents an *immediately contains* relation between a corresponding pair of nested regions.) In the remainder we will often think of partitions as such planar maps (that is as Venn diagrams without overlapping), and the minimal cells correspond to the smallest regions within such diagrams.

Tree and Venn-diagram representations of granular partitions are not equivalent. To see this consider Figure 2. Even if we ignore the labeling it is obvious that the two Venn-diagrams represent two distinct partitions. The mammal-partition contains 'empty space' and the first-couple-partition is full in the sense that it does not contain 'empty space'. This distinction, however, can not be made in terms of the corresponding tree representations. In order to represent it in the tree we needed consider labeled trees with nodes labeled *full* or *not-full*. We will discuss these issues in more detail in our section on fullness and cumulativeness of granular partitions.

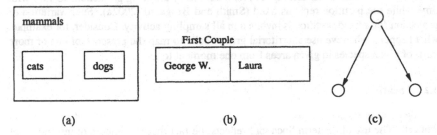

Figure 2: Venn-diagram and tree representations of granular partitions

4 GRANULAR PARTITIONS IN THEIR PROJECTIVE RELATION TO REALITY

4.1 Projection

Partitions are more than just systems of cells. They are constructed to serve as inventories or pictures or maps of specific portions of reality, and in this they are analogous to *windows*, or to the latticed grills purported to have been used by Renaissance artists as aids to the faithful representation of objects in reality (Smith, 2001b). They are analogous also to propositions (*Elementarsätze*) as described by Wittgenstein in the *Tractatus* (1961). A proposition, for Wittgenstein, is built out of simple signs (names) arranged in a certain order. Each name, Wittgenstein tells us, stands in a projective relation to a corresponding object in the world: it cannot fail to strike its target. If a proposition is true, then its simple signs stand to each other within the proposition as the corresponding objects stand to each other in the world. It is in this sense that a true atomic proposition is a picture, as Wittgenstein puts it, of a state of affairs in reality. That a proposition is a complex of names arranged in a certain order is in our present context equivalent to the thesis that a partition is a complex of cells arranged in a certain order.

A partition is a complex of cells in its projective relation to the world (compare *Tractatus*, 3.12). This relation may be effected either directly by the user of the partition—for example in looking through the cells of the grid and recording what objects are detected on the other side—or indirectly, with the help of proper names or other referring devices such as systems of coordinates or taxonomic labels.

For Wittgenstein it is guaranteed a priori for every name that there is some unique object onto which the name is projected. From the perspective of the theory of granular partitions, in contrast, projection may fail. That is, a partition may be such that—like the partition cataloguing Aztec gods—there are no objects for its cells to project onto. Works of fiction and also not yet realized plans may be conceived as involving partitions of this kind.

In this paper, however, we are interested primarily in partitions which do not project out into thin air in this way. We write '$P(z, o)$' as an abbreviation for: cell z is projected onto object o. We can also, if the context requires it, write '$P_A(z, o)$' to indicate that the projection of z onto o obtains in the context of partition A. In what follows we shall assume that a unique such projection is defined for each partition. In a more general theory we can weaken this assumption, for example by allowing projections to vary with time while the partition remains fixed (Smith and Brogaard, 2002a). Such variation of projection for a fixed partition is involved in all sampling activity. Consider, for example, what happens when we use a territorial grid of cells to map the presence of one or more birds of given species in given areas from one moment to the next.

4.2 Location

If projection is successful, then the object upon which a cell is projected is located in that cell. The use of the term 'location' reflects the fact that one important inspiration of our work is the study of location relations in spatial contexts. One motivating example of a location relation within our theory is the relation between a spatial object such as a railway station and an icon on a map. Other motivating examples are of a non-spatial sort: they include the relation between an instance (Tibbles) and its kind (cat) or the relation between a customer and the corresponding record in a database. Indeed they include the relation between you and your name.

We can compare a partition with a rig of spotlights projecting down onto an orchestra during the performance of a symphony. Each cell of the partition corresponds to some spotlight in the rig. Some cells (spotlights) will project upon single players, others onto whole sections of the orchestra (string, wind, percussion, and so forth). One cell (spotlight) will project upon the orchestra as a whole. Note that the spotlights do not hereby create the objects which they cast into relief. When once the rig has been set, and the members of the orchestra have taken their places, then it will be an entirely objective matter which objects (individuals and groups of individuals) are located in which illuminated cells.

In what follows we make the simplifying assumption that objects are exactly located at their cells (that spotlights never partially illuminate single players or sections). Compare the way in which Wyoming is exactly located at the cell 'Wyoming' in the partition of the US into States or the way in which your brother Norse is exactly located at the cell 'Norse' in your partition (list) of your family members. In a more general theory we liberalize the location relation in such a way as to allow also for partial or rough location (Casati and Varzi, 1995; Bittner and Stell, 1998).

4.3 Transparency

When projection succeeds, then location is what results. Projection and location thus correspond to the two directions of fit—from mind to world and from world to mind—between an assertion and the corresponding truthmaking portion of reality (Searle, 1983; Smith, 1999). Projection is like the relation which holds between your shopping list and the items which, if your shopping trip is successful, you will actually buy. Location is like the relation which obtains between the items you have bought and the new list your mother makes after your return, as she checks off those items which you have in fact succeeded in bringing back with you.

The formula '$L(o, z)$' abbreviates: object o is located at cell z. (And again where this is required we can write '$L_A(o, z)$' for: o is located at z in partition A.) Location presupposes projection: an object is never located in a cell unless the object has already been picked out as the target of the projection relation associated with the relevant partition. But *successful* projection—by which is meant the obtaining of the projection relation between a cell and an object—also presupposes location, so that where both L and P obtain they are simply the converse relations of each other. We have now reached the point where we can formulate the first of our master conditions on partitions from the perspective of theory (B):

MB1: $L(o, z) \rightarrow P(z, o)$
MB2: $P(z, o) \rightarrow L(o, z)$

(Successful) projection and (successful) location are simple converses of each other. (We formulate this principle as two separate conditions in order to leave room for a more general theory in which these two relations are teased apart.)

MB1 and MB2 tell us that a partition projects a given cell onto a given object if and only if that object is located in the corresponding cell. Very many partitions—from automobile component catalogues to our maps of states and nations—have this quality without further ado.

We shall call partitions which satisfy MB1 and MB2 *transparent* partitions, a notion which we can define in the obvious way as follows:

DTr: $Tr(A) \equiv \forall z \forall o : P_A(z, o) \leftrightarrow L_A(o, z)$

MB1 and MB2 jointly ensure that objects are actually located at the cells that project onto them. Notice however that a transparent partition, according to our definition, may still have empty cells. Such cells may for example be needed in the context of scientific partitions in order to leave room for what, on the side of the objects, may be discovered in the future. (Compare the cells labeled Ununnilium, Unununium and Ununbium in the Periodic Table of the Elements.) Empty cells may similarly be needed to cover up for temporary lapses in memory. You are attempting to account for the people at your party last night. Your partition consists of six cells labeled: *John, Mary, Phil, Chris, Sally,* and *anyone else* (for people you might have forgotten). Assume that John, Mary, Phil, Chris and Sally is a complete listing of all the people at the party. Your *anyone else* cell is then empty.

4.4 Functionality constraints (constraints pertaining to correspondence to objects)

4.4.1 Projection is functional: The confused schoolboy

The property of transparency is still rather weak. Thus transparency is consistent with ambiguity on the side of the cells in relation to the objects they target, that is with the case where one cell projects onto two distinct objects. An example of the sort of problem we have in mind is the partition created by a lazy schoolboy studying the history of the Civil War in England. This partition has one cell labeled 'Cromwell'—and so it does not distinguish between Oliver and his son Richard. Another example might be the partition utilized by those who talk of 'China' as if the Republic of China and the People's Republic of China were one single object.

To eliminate such ambiguity we lay down a requirement to the effect that each partition must be such that its associated projection is a *functional* relation:

MB3: $P(z, o_1)$ and $P(z, o_2) \rightarrow o_1 = o_2$

For partitions satisfying MB3, cells are projected onto single objects (one rather than two). Consider the left part of Figure 3. The dotted arrow can occur in partitions satisfying merely MB1–2 but not in partitions also satisfying MB3. Notice, though, that projection might still be a partial function, since MB3 does not rule out the case where there are empty cells.

To impose the functionality of projection on all partitions is in one respect trivial. For we can very easily convert a partition A which does not satisfy MB3 into one which does. If z is a cell in A which does not satisfy MB3, then we create this new partition A' by adjusting A in such a way that z now projects upon the mereological sum of the objects

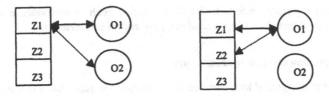

Figure 3: Transparent partitions in which projection is not functional (left); location is not functional (right)

its projects upon in A. This account seems, indeed, to do justice to what is involved in the confused schoolboy case, namely that Richard and Oliver are run together, somehow, into one composite human being.

In the remainder of this paper we use the notation $o = p(z)$ instead of $P(z, o)$ whenever we assume that projection is functional.

4.4.2 Location is functional: The Morning Star and the Evening Star

Consider a partition having root cell labeled 'heavenly bodies' and three subcells labeled: 'The Morning Star', 'The Evening Star', and 'Venus', respectively. As we know, all three subcells project onto the same object. This partition is perfectly consistent with the conditions we have laid out thus far. Its distinct subcells truly, though unknowingly, project onto the same object. It is not unusual that we give different names (or class-labels) to things in cases where we do not know that they are actually the same. A good partition, though, should clearly be one in which such errors are avoided.

Partitions manifesting the desired degree of correspondence to objects in this respect must in other words be ones in which location, too, is a *functional* relation:

MB4: $L(o, z_1)$ and $L(o, z_2) \rightarrow z_1 = z_2$

In partitions that satisfy MB4, location is a function, i.e., objects are located at single cells (one rather than two). Consider the right part of Figure 3. The dotted arrow can occur in partitions satisfying MB1–2, not however in partitions also satisfying MB4. As MB3 rules out co-location (overcrowding), so MB4 rules out co-projection (redundancy). Note that natural analogues of co-location and co-projection are not even formulable within a set-theoretic framework.

5   CORRESPONDENCE OF MEREOLOGICAL STRUCTURE

MB1 and MB2 are, even when taken together with MB3 and MB4, still very weak. They tell us only that, if a cell in a partition projects upon some object, then that object is indeed located in the corresponding cell. They do not tell us what happens in case a cell fails to project onto anything at all. MB1–4 thus represent only a first step along the way towards an account of correspondence to reality for partitions. Such correspondence will involve the two further dimensions of *structural mapping* and of *completeness*.

5.1 Recognizing mereological structure

An object o is *recognized* by a partition if and only if the latter has a cell in which that object is located (Smith and Brogaard, 2002b). Intuitively, recognition is the partition-theoretic analogue of the standard set-membership relation. Partitions embody the selective focus of our mapping, classifying, and listing activities. To impose a partition on a given domain of reality is to *foreground* certain objects and features in that domain and trace over others. Note hereby that we trace over not only the objects which *surround* that which is foregrounded, as according to the usual understanding of the fore-ground/background structure; for we also trace over those *parts* of the foregrounded object which fall beneath the threshold of our concerns. Partitions are granular in virtue precisely of the fact that a partition can recognize an object without recognizing all its parts.

Partitions—think again of Venn diagrams—are designed to reflect the part-whole structure of reality through the fact that the cells in a partition are themselves such as to stand in relations of part to whole. Given the master conditions expressed within the framework of theory (A) above, partitions have at least the potential to reflect the mereological structure of the domain onto which they are projected. And in felicitous cases this potential is realized.

That we distinguish between the recognition (foregrounding, selection) of objects on the one hand and the reflection of mereological structure on the other hand is not an arbitrary matter. In Tractarian semantics we distinguish between projection and isomorphism. In set theory we distinguish, for any given set, between a domain of elements and the set-theoretic structure imposed on this domain. Just as it is possible to have sets consisting entirely of *Urelemente* (together with a minimal amount of set-theoretic packaging), so it is possible to have partitions built exclusively out of minimal cells (and one root cell). Such partitions amount, simply, to lists of the things that are recognized by their cells, with no mereological structure on the side of these objects being brought into account.

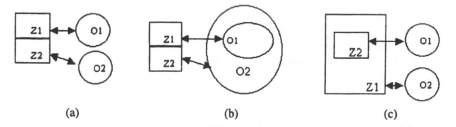

(a)   (b)   (c)

Figure 4: Transparent partitions with more or less desirable properties

Figure 4(a) and 4(b) represent partitions consisting of two minimal cells z_1 and z_2 projecting onto objects o_1 and o_2. Case (a), a simple list, is unproblematic. Case (b) we shall also allow. This is in keeping with the notion that minimal cells are the (relative) atoms of our system, and we take this to mean that they should be neutral with regard to any mereological structure on the side of their objects. An example of type (b) would be a list of regions represented at a conference to discuss measures against terrorism, a conference including representatives from both Germany and Bavaria.

Cases like (c), in contrast, represents projections in which, intuitively, something has gone wrong. All three cases satisfy the master conditions we have laid down thus far, for the latter allow both for disjoint cells to be projected onto what is not disjoint (b) and also for disjoint objects to be located in cells which are not disjoint (c). Cases like (c) on the other hand seem to fly in the face of a fundamental principle underlying the practice of hierarchical classification, namely that objects recognized by species lower down in a hierarchical tree should be included as parts in whatever is recognized by the genera further up the tree. To exclude cases like (c) we shall impose a condition to the effect that mereological structure within a partition should *not misrepresent the mereological relationships* between the objects which the corresponding cells are projected onto. We first of all define the following relation of *representation of mereological structure* between pairs of cells:

DS1: $RS(z_1, z_2) \equiv \forall o_1, o_2 : (L(o_1, z_1) \text{ and } L(o_2, z_2) \text{ and } z_1 \subseteq z_2) \rightarrow o_1 \leq o_2$

If z_1 is a subcell of z_2 then any object recognized by z_1 must be a part of any object recognized by z_2. A partition is then mereologically structure-preserving if and only if each pair of cells within the partition satisfies DS1:

DS2: $RS(A) \equiv \forall z_1, z_2 : (Z(z_1, A) \text{ and } Z(z_2, A)) \rightarrow RS(z_1, z_2).$

We can now impose a new master condition:

MB5: All partitions are *mereologically structure-preserving* in the sense of DS2.

Note that even MB5 is still very weak. Its effect is entirely negative, since it merely ensures that partitions do not misrepresent the mereological relationships between their objects. Partitions might still be entirely blind to (trace over) such relationships. Two minimal cells might project onto objects which stand to each other in any one of the possible mereological relations (identity, proper parthood, disjointness, overlap), and all pairs of cells are likewise neutral as to the mereological relations between the objects onto which they are projected provided only that they do not stand to each other in the subcell relation. This means that, given such cells, we are entitled to infer nothing at all about the mereological relations among the corresponding objects.

Consider, for example, a partition that contains cells, z_1 and z_2, that recognize John and his arm, respectively, so that $L(John, z_1)$ and $L(John's\ arm, z_2)$. Then cell z_1 need not be a proper subcell of the cell z_2, for the partition may not know that the object located in z_2 is properly designated as John's arm. Or consider a partition containing two cells that recognize, respectively, mammals and whales. Suppose that this is a partition constructed at a time when the status of whales as mammals was not yet recognized. The cell labeled *whales* is not, then, included as a subcell of the cell labeled *mammals*. But the partition can still satisfy our conditions laid down so far. This is so, for example, if the cell that recognizes whales is a subcell of the cell recognizing animals but not a subcell of any other subcell of the cell recognizing animals (Partition A_1 in Figure 5). If the cell that recognizes whales were also a subcell of the cell that recognizes fish, for example, then the partition would misrepresent the mereological relationship between these two species and so violate MB5 (Partition A_2 in Figure 5).

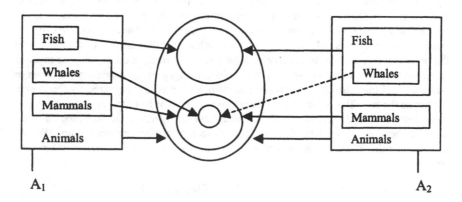

Figure 5: Partition A_1 does not misrepresent the mereological structure of the underlying domain. Partition A_2 places whales incorrectly in relation to fish and mammals

Partitions may trace over mereological relationships between the objects they recognize, but MB5 is strong enough to ensure that, if a partition tells us something about the mereological relationships on the side of the objects which it recognizes, then what it tells us is true. Notice that partition A_2 still satisfies MB1–4.

Consider a domain of objects consisting of two regions, x and y, that properly overlap in the region v, so that $x * y = v$ with $v < x$ and $v < y$. Consider now a partition with cells z_1 and z_2 recognizing x and y, respectively, so that $L(x, z_1)$ and $L(y, z_2)$. Assume further that z_1 and z_2 do not stand in any subcell relation to each other, i.e., their partition-theoretic intersection is empty. Only four possibilities regarding the representation of v now remain: (1) our partition does not recognize v at all; (2) it recognizes v but traces over its mereological relationships to x and y; (3) it recognizes v through a subcell of z_1 but it traces over the fact that v could equally well be recognized by a subcell of z_2; (4) it recognizes v through a subcell of z_2 but it traces over the fact that v could equally well be recognized by a subcell of z_1. The fifth possibility—of allowing sub-cells of both z_1 and z_2 to recognize v is excluded by the tree structure of granular partitions.

Let x and y be two neighboring countries which disagree about the exact location of their common boundary and let v be the disputed area. The inhabitants of country x consider v to be part of x, the inhabitants of country y consider v to be part of y. Possibility (1) then corresponds to the view of some third country at the other side of the globe who recognizes the countries x and y but does not care about their border dispute. (2) corresponds to the view of an observer who recognizes that there is a disputed area but who is neutral about the status of the disputed area. (3) corresponds to the view of country x and (4) to that of country y.

Another example of case (2) is provided by Germany and Luxemburg, which overlap at their common border on the River Our. The river is part of both countries. Mapmakers normally have no facility to represent cases such as this, and so they either adopt the policy of not representing such common regions at all (the border is represented as a line which we are to imagine as being without thickness), or they recognize the region constituted by the river on the map but trace over its mereological properties. Larger-scale maps often embrace a third alternative, which is to *misrepresent* the relations between Germany and Luxemburg by drawing the boundary between the two countries as running down the center of the river.

5.2 The domain of a partition

That upon which a partition is projected is a certain domain of objects in reality (the term 'domain' being understood in the mereological sense). We shall conceive the domain of a partition as the mereological sum of the pertinent objects. It is, as it were, the total mass of stuff upon which the partition sets to work: thus it is stuff conceived as it is prior to any of the divisions or demarcations effected by the partition itself. The domains of partitions will comprehend not only individual objects and their constituents (atoms, molecules, limbs, organs), but also groups or populations of individuals (for example biological species and genera, battalions and divisions, archipelagos and diasporas) as well as their constituent parts or members. Domains can comprehend also extended regions (continua) of various types. Spatial partitions, for example maps of land use or soil type (Frank et al., 1997), are an important family of partitions with domains of this sort. There are also cases where partitions impose upon continuous domains a division into discrete units for example by creating temperature or frequency bands.

We are now able to specify what we mean by 'domain of a partition.' Our representation of partitions as trees and our condition on reflection of structure (MB5) ensure that all partitions trivially reflect the fact that the objects recognized by their cells are parts of some mereological sum. For MB5 is already strong enough to ensure that everything that is located at some cell of a partition is part of what is located at the corresponding root cell. If any cell pointed outside of what is located at the root cell it would misrepresent the mereological structure of the corresponding domain.

We can thus define the domain of a partition simply as the object (mereological whole) onto which its root cell is projected. By functionality of projection and location there can be only one such object.

DD: $D(A) = p(r(A))$

We now demand as a further master condition that every partition has a non-empty domain in the sense of DD:

MB6: $\exists x : x = D(A)$

We then say that a partition *represents its domain correctly* if and only if MA1–5 and MB1–6 hold. Note that this condition of *correctness* is still rather easily satisfied. (It is

achieved already in every simple list, provided only that the list involves no double counting and no ambiguous reference of the sort involved in the Oliver and Richard Cromwell
case.)

If there is a single maximal object (the whole universe), then one correct representation thereof is provided by a partition consisting of just one cell labeled 'everything' (we
might call this the Spinoza partition). A partition with just three cells: a root cell, labeled
animals, and two subcells, labeled *dogs* and *cats*, represents its domain correctly; it just
falls far short of a certain desirable completeness. Correct representations, as we see, can
be highly partial.

5.3 The granularity of granular partitions

A correct representation, as we see, is not necessarily a complete representation. Indeed,
since partitions are cognitive devices, and cognition is not omniscient, it follows that no
partition is such as to recognize all objects. There is no map of all the objects in the
universe. The complexity of the universe is much greater than the complexity of any
single cognitive artifact. This feature of partiality is captured already by our terminology
of *granular* partitions. Partitions characteristically do not recognize the proper parts of
the whole objects which they recognize; for example they do not recognize parts which
fall beneath a certain size.

It is the cells of a partition which carry with them this feature of granularity. Because
they function like singletons in set theory, they recognize only single whole units, the
counterparts of set-theoretic elements or members. If a partition recognizes not only
wholes but also one or more parts of such wholes, then this is because there are additional
cells in the partition which do this recognizing job. Consider, for example, a partition
that recognizes human beings and has cells that project onto John, Mary, and so forth.
This partition does not recognize parts of human beings—such as John's arm or Mary's
shoulder—unless we add extra cells for this purpose. Even if a partition recognizes both
wholes and also some of their parts, it is not necessarily the case that it also reflects the
mereological relationships between the two. Imagine we are forensic scientists examining
photographs taken at a crime scene and that these photographs generate a partition with
cells recognizing John, Mary, and an arm. It may then be the case that the state of our
knowledge is such that the cell recognizing the arm is not a subcell of the cell recognizing
either John or Mary. Or let the arm be Kashmir and let John and Mary be India and
Pakistan, respectively.

In relation to this granularity of partitions, we can once more call in the aid of Wittgenstein:

> In the proposition there must be exactly as many things distinguishable as
> there are in the state of affairs, which it represents. They must both possess
> the same logical (mathematical) multiplicity ... (4.04)

Wittgenstein himself takes care of the issue of granularity by insisting that the world is
made up of discrete simples, and by insisting further that all partitions (for Wittgenstein:
propositions) picture complexes of such simples. (A similar simplifying assumption is
proposed by Galton, 1999.) This is a simplifying assumption, which our present theory

of granular partitions will enable us to avoid. For the latter admits partitions of arbitrary granularity including partitions which reflect distinct cross-cuttings of the same domain of reality (Smith and Brogaard, 2002a). The theory of granular partitions enables us moreover to remain neutral as to the existence of any ultimate simples in reality from out of which all other objects would be constructed via summation. This is due to the fact that partitions are by definition *top-down* structures. The duality with trees puts special emphasis on this aspect: we trace down from the root until we reach a leaf. A leaf has no further parts within the partition to which it belongs. But it need not necessarily project upon something that itself has no parts. The fact that there are leaves simply indicates that a partition does not care about (traces over) what lies beneath a certain level of granularity on the side of its objects. An object located at a minimal cell is an atom only relative to the partition involved.

Partitions are cognitive devices which have the built-in capability to recognize objects and to reflect certain features of the latters' mereological structure and to ignore (trace over) other features of this structure. We can now see that they can perform this task of tracing over in two ways: (1) by tracing over mereological relations between the objects which they recognize; (2) by tracing over (which means failing to recognize) parts of those objects. (2) is (unless atomism is true) a variety of tracing over that must be manifested by every partition. A third type of tracing over arises in reflection of the fact that partitions (we leave to one side here the Spinoza partition) are partial in their focus. In foregrounding some regions of reality each partition thereby traces over everything that lies outside its domain.

Consider a simple biological partition of the animal kingdom including a cell projecting on the species dog (*Canis familiaris*). Our definition of the domain of a partition and our constraint on functionality of projection implies that, besides the species dog also your dog Fido, and also Fido's DNA-molecules, proteins, and atoms are parts of the domain of this partition. *But the latter are of course not recognized by the partition itself.* It is cases such as this which illustrate why mereology requires supplementation by a theory like the one presented here. Partition theory allows us to define a new, restricted notion of parthood that takes granularity into account (compare Degen et al., 2001). This restricted parthood relation is an analogue of partition-theoretic inclusion, but on the side of objects:

$$\text{DRP:} \quad x \leq_A y \equiv \exists z_1, z_2 : L_A(x, z_1) \text{ and } L_A(y, z_2) \text{ and } z_1 \subseteq z_2$$

This means that x is a part of y relative to partition A if and only if: x is recognized by a subcell of a cell in A which recognizes y. From this we can infer by MB5 that x is a part of y also in the unrestricted or absolute sense.

The usual common-sense (i.e., non-scientific) partition of the animal kingdom contains cells recognizing dogs and mammals, but no cells recognizing DNA molecules. Relative to this common-sense partition, DNA molecules are not parts of the animal kingdom in the sense defined by DRP, though they are of course parts of the animal kingdom in the usual, non-relativised sense of 'part'.

6 STRUCTURAL PROPERTIES OF CORRECT REPRESENTATIONS

In this section we discuss some of the more fundamental varieties of those partitions which satisfy the master conditions set forth above. We classify such partitions according to: (1) degree of structural fit; (2) degree of completeness and exhaustiveness; (3) degree of redundancy.

6.1 Mereological monotony

We required of partitions that they at least not misrepresent the mereological structure of the domain they recognize. This constraint is to be understood in such a way that it leaves room for the possibility that a partition is merely neutral about (traces over) some or all aspects of the mereological structure of its target domain. Taking this into account, we can order partitions according to the degree to which they actually do represent the mereological structure on the side of the objects onto which they are projected. At the maximum degree of structural fit we have those partitions which completely reflect the mereological relations holding between the objects which they recognize.

Such a partition satisfies a condition to the effect that if o_1 is part of o_2, and if both o_1 and o_2 are recognized by the partition, then the cell at which o_1 is located is a subcell of the cell at which o_2 is located. Such partitions satisfy the weak converse of MB5. Formally we can express this constraint on mereological structure (CM) as follows:

CM: $L(o_1, z_1)$ and $L(o_2, z_2)$ and $o_1 \leq o_2 \rightarrow z_1 \subseteq z_2$

A partition satisfying CM is *mereologically monotonic*. This means that it recognizes all the restricted parthood relations obtaining in the pertinent domain of objects. A very simple example is given by a flat list (a partition having only minimal cells together with a root) projected one-for-one upon a collection of disjoint objects.

6.2 Completeness

So far we have allowed partitions to contain empty cells, i.e., cells that do not project onto any object. We now consider partitions which satisfy the constraint that every cell recognizes some object:

CC: $Z(z, A) \rightarrow \exists o : L(o, z)$

We say that partitions that satisfy CC *project completely*. Notice that this condition is independent of the functional or relational character of projection and location. Of particular interest, however, are partitions that project completely and in such a way that projection is a total function (partitions which satisfy both MB3 and CC). An example is a map of the United States representing its constituent states. There are no no-man's lands within the territory projected by such a map and every cell projects uniquely onto just one state.

6.3 Exhaustiveness

So far we have accepted that there may be objects in our target domain that are not located at any cell. This feature of partitions is sometimes not acceptable: governments want *all* their subjects to be located in some cell of their partition of taxable individuals. They want their partitions to satisfy a completeness constraint to the effect that every object in the domain is indeed recognized. In this case we say that location is *complete*. Alternatively we say that the partition *exhausts* its domain. Unfortunately, we cannot use

(∗) $o \leq D(A) \rightarrow \exists z : Z(z, A)$ and $L(o, z)$

to capture the desired constraint. The tax authorities do not (as of this writing) want to tax the separate molecules of their subjects. Trivially, we have:

$o \leq_A D(A) \rightarrow \exists z : Z(z, A)$ and $L(o, z)$

but this is much too weak, since it asserts only that every object within a given domain that is recognized by a partition is indeed recognized by that partition. It will in fact be necessary to formulate several restricted forms of exhaustiveness, each one of which will approximate in different ways to the (unrealizable) condition expressed in (∗).

One such exhaustiveness condition might utilize a sortal predicate (schema) φ that singles out the kinds of objects our partition is supposed to recognize (for example, in the case of the partition of taxable individual human beings, rather than proper parts of human beings). We now demand that the partition A recognize all of those objects in its domain which satisfy φ:

CE_φ: $o \leq D(A)$ and $\varphi(o) \rightarrow \exists z : Z(z, A)$ and $L(o, z)$

Let Δ be some domain and let A be a partition such that $\Delta = D(A)$. Since we can very simply use any predicate to define a partition over any domain—by setting

$L_A(o, z) \equiv o < \Delta$ and $\varphi(o)$

—we can also think of CE_φ as asserting the completeness of one partition *relative to* another. Note that the idea underlying CE_φ is closely related to the idea of granularity. Thus for some purposes we might find it useful to formulate condition φ as a restriction on object size.

The tax office probably does not care too much about empty cells in its partition, nor is it bothered too much by the idea of charging you twice. The main issue is to catch everything above a certain resolution at least once. This is the intuition behind constraints like CE_φ. If you are a law-abiding citizen, you will accept CE_φ (where 'φ' stands for 'is a citizen'), but you will insist that the partition not locate you in two separate cells, i.e., that you are not charged twice. This means that you want the tax partition to satisfy CE_φ and MB4. There might be a pedantic clerk in the tax office who does not rest until he has made sure that all empty cells have been removed. Partitions that will satisfy you, the government, and the clerk in the tax office must satisfy CC, CE_ϕ, and MB1–5. Projection and location are then total functions (relative to a selected predicate ϕ) and one is the inverse of the other. Under those circumstances projection and location are bijective functions. Notice that neither of the following holds:

(**) if MB4 and CE_ϕ and CC then MB3
(* * *) if MB3 and CE_ϕ and CC then MB4

Counterexamples are given in Figure 6 (a) and (b), respectively, where each depicted object is assumed to satisfy φ.

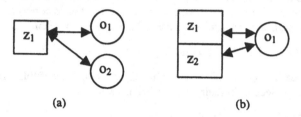

(a)   (b)

Figure 6: Functionality of projection and location are independent of completeness and exhaustiveness

6.4 Comprehension axioms

The following is the partition-theoretic equivalent of the unrestricted set-theoretic comprehension axiom. For each predicate φ there is a partition $A(\varphi)$ whose location relation is defined as follows:

$$\exists z : L_{A(\varphi)}(o, z) \text{ iff } \varphi(o)$$

Under what conditions on φ can this be allowed?

One type of restriction that is relevant to our purposes would allow φ to be unrestricted but affirm additional restrictions on objects, for example in terms of spatial location. Thus we might define a family of spatial partitions $A(\varphi, r)$, where r is some pre-designated spatial region, in such a way that

$$\exists z : L_{A(\varphi,r)}(o, z) \text{ iff } \varphi(o) \text{ and } o \text{ is spatially located in } r.$$

Something like this is in fact at work in the taxation partition (the tax office is interested in human beings bearing a special relation to a specific geographic location), as also in the partitions used by epidemiologists, ornithologists and others who are interested in (types of) objects at specific sites.

6.5 Redundancy

Partitions are natural cognitive devices, for example they are lists, maps, and so forth, used by human beings to serve various practical purposes. This means that partitions will normally be called upon to avoid certain sorts of redundancy. Here we distinguish what we shall call correspondence redundancy and structural redundancy. Necessarily empty cells (cells whose labels tell us *ex ante* that no objects can be located within them) represent one type of correspondence redundancy, which is excluded by condition CC.

Another type of correspondence redundancy we have addressed already in our discussion of the functionality of location. This occurs in a partition with two distinct cells whose labels would tell us, again *ex ante*, that they must necessarily project upon the very same object. Clearly, and most simply, a partition should not contain two distinct cells with identical labels.

The following case is not quite so trivial. Consider a partition with a cell labeled *vertebrates* which occurs as a subcell of the cell labeled *chordates* in our standard biological classification of the animal kingdom. Almost all chordates are in fact vertebrates. Suppose (for the sake of argument) that biologists were to discover that all chordates must be vertebrates. Then such a discovery would imply that, in order to avoid structural redundancy, they would need to collapse into one cell the two cells (of chordates and vertebrates) which at present occupy distinct levels within their zoological partitions.

A constraint designed to rule out such structural redundancy would be:

CR: A cell in a partition never has exactly one immediate descendant.

This rules out partition-theoretic analogues of the set theorist's $\{\{a\}\}$.

7 FULLNESS AND CUMULATIVENESS

There are a number of different sorts of knowledge shortfall which should be considered in any complete theory of granular partitions. One type arises when there are missing levels within a hierarchy. A partition of the United Kingdom which mentions regions, counties, towns, etc., but leaves out the cells *England, Scotland, Wales* and *Northern Ireland* is an example of this sort of incompleteness.

More important however are those types of shortfall which have to do with relations between existing levels. We have distinguished thus far *completeness*, which has to do with the absence of empty cells, and *exhaustiveness*, which has to do with the successful capturing of all pertinent objects in a given domain. But shortfalls arise also in relation to a third type of completeness, which has to do with ensuring that successive levels of a partition relate to each other in the most desirable way. We can initially divide this third type of completeness into two sub-types: fullness and cumulativeness. Fullness, intuitively, is a requirement to the effect that each cell z has enough daughter (immediate descendant) cells to fill out z itself. Cumulativeness is a requirement to the effect that these daughter cells are such that the objects onto which they are projected are sufficient to exhaust the domain onto which the mother cell is projected. Fullness, accordingly, pertains to theory (A), cumulativeness to theory (B). (We shall henceforth assume for the sake of simplicity that there are no redundancies in the sense of CR.)

Non-fullness and non-cumulativeness represent two kinds of shortfall in the *knowledge* that is embodied in a partition. Non-fullness is the shortfall which arises when a cell has insufficiently many subcells within a given partition (for instance it has a cell labeled *mammal*, but no subcells corresponding to many of the species of this genus). Non-cumulativeness is the shortfall which arises when our projection relation locates insufficiently many objects in the cells of our partition, for example when I strive to make a list of the people that I met at the party yesterday but leave out all the Welshmen. Fullness and cumulativeness are rarely satisfied by our scientific partitions of the natural world.

They are satisfied primarily by artificial partitions of the sort which are constructed in database environments.

7.1 Fullness

Consider a partition consisting of three cells, labeled *people*, *Laura* and *George W.* Or consider a partition with three cells labeled: *mammals*, *cats* and *dogs*. Both of these partitions are transparent, by our definition (DTr); but both are, again, such as to fall short of a certain sort of ideal completeness, which we can express by asserting that the mereological sum of the cells *Laura* and *George W.* (or of the cells *cats* and *dogs*) falls far short of the corresponding partition-theoretic sum.

If a collection of subsets of some given set forms a partition of this set in the standard mathematical sense, then these subsets are (1) mutually exhaustive and (2) pairwise disjoint. An analogue of condition (2) holds for minimal cells in our present framework, since minimal cells are always mereologically disjoint (they cannot, by definition, have subcells in common). Condition (1) however does not necessarily hold within the framework of partition theory. This is because, even where the partition-theoretic sum of minimal cells is identical to the root cell, the minimal cells still do not necessarily exhaust the partition as a whole. The mereological sum (+) of cells is, we will recall, in general smaller than their partition-theoretic sum ($\cup$).

We call a cell *full* relative to its descendant cells within a given partition if these descendants are such that their mereological sum and their partition-theoretic sum coincide. Formally we define:

DFullcell: $Fullcell(z_1) \equiv (+_{z \sqsubset z_1} z) = (\cup_{z \sqsubset z_1} z)$

where $(+_{z \sqsubset z_1} z)$ and $(\cup_{z \sqsubset z_1} z)$ symbolize respectively the operations of applying mereological and partition-theoretic sum for all proper subcells z of the cell z_1. Since we can easily prove that $(\cup_{z \sqsubset z_1} z) = z_1$, DFullcell could be reformulated as asserting that a cell is full if and only if it is identical to the mereological sum of its descendants:

$$Fullcell(z_1) \leftrightarrow +_{z \sqsubset z_1} z = z_1$$

DFullcell does not suffice to capture the intended notion of fullness for partitions however. To see the problem, consider the partition consisting of

top row: one maximal cell, labeled *first couple*
middle row: two intermediate cells, labeled *George W.* and *Laura*
bottom row: four minimal cells, labeled *George W.'s left arm*, *George W.'s right leg*, *Laura's left arm*, *Laura's right leg*.

The cell in the top row satisfies DFullcell, i.e. it is full relative to the second row; but it is not full relative to all of its descendants, since the mereological sum of the cells *George W.'s arm* and *George W.'s leg* is not identical to the cell *George W.* (and analogously for *Laura*). The problem arises because if $x \leq y$ then $x + y = y$ and if $x \subseteq y$ then $x \cup y = y$. From this it follows that only the immediate descendants of a given cell z_1 contribute to its mereological and partition-theoretic sums.

This, however, tells us what we need to take into account in defining what it is for a cell to be full relative to all its descendant cells within a given partition A, namely that each of its constituent cells must be full relative to its immediate descendents. This yields:

DFullcell*: $Fullcell*(z_1, A) \equiv \forall z : z \subseteq z_1 \rightarrow (Fullcell(z)$ or $Min(z, A))$

Here minimal cells have been handled separately because they do not have subcells. One can see that, while *Fullcell*(*first couple*) holds in the mentioned partition, *Fullcell**(*first couple*) does not, because the cells *George W.* and *Laura* are neither full nor minimal.

We can now define what it means for a *partition* to be full, as follows:

DFull2: $Full(A) \equiv \forall z : Z(z, A) \rightarrow (Fullcell(z)$ or $Min(z, A))$

A partition is full if and only if all its non-minimal cells are full (or, equivalently, all its non-minimal cells are full*). Notice that full partitions might in principle contain empty cells, which may or may not have subcells. (Consider, for example, the cell *dodo* with subcells *male dodo* and *female dodo*.)

7.2 Empty space

When a cell falls short of fullness, this means that, while the cell successfully projects onto some given domain, its subcells do not succeed in projecting onto the entirety of this domain. It is then as if there is some extra but invisible component in the cell, in addition to its subcells. We shall call this additional component 'empty space' (noting that the term 'empty' here has a quite different meaning from what it has in the phrase 'empty cell'). Consider the partition depicted in Figure 2.

Here the empty space is that part of the cell *mammals* that is not occupied by *cats* and *dogs*. Notice that this empty space is a component of the cell *mammals* but it is not itself a cell nor is it made up of cells. Empty space is that part of a cell that is not covered by its subcells. The notion of empty space is then similar to the notion of hole in the sense of Casati and Varzi (1994). A hole requires an object which serves as its host. A hole is in every case a hole *in* something. In the same sense empty space requires a cell within which it can exist as empty space. As a hole is a concavity of a host object where no parts of this host object are to be found, so empty space is a zone within a cell where no subcells are to be found.

Empty space within a cell is like a hole also in the sense that there must be something that potentially fills it. In our case this means: more subcells. If all the empty space in a cell is filled, then the cell itself is full and the empty space has been eliminated. Empty space is inert in the sense that it does not project onto anything. Empty space is normally hidden to the user of the partition in which it exists, for otherwise this user would surely have constructed a fuller partition. In some cases however a user might deliberately accept empty space in order to have means to acknowledge the fact that something has been left out. Alternatively, the existence of empty space in a given partition might be brought to the attention of the user. We point in a certain direction and ask: What is *there*? The theory of empty space thereby serves as the starting-point for an ontology of questions (Schuhmann and Smith, 1987): empty space corresponds to a hole in our knowledge.

7.3 Empty space and knowledge

The presence or absence of empty space is a dimension of a granular partition that is skew to the dimension pertaining to the existence of empty cells. An empty cell is a cell that fails to project. Empty space is that which leaves room for the addition of new knowledge. It is a zone within a cell that is not (but given advances in knowledge, could be) occupied by further subcells. Notice, however, that there cannot be a partition that is full and yet consists entirely of empty cells. This is because by MB6 every partition projects onto some object.

Figure 2 (a) depicts a partition of the animal kingdom consisting of three cells, where z_3 recognizes the animal kingdom as a whole, z_1 recognizes dogs, and z_2 recognizes cats. In terms of partition-theoretic union we have $z_1 \cup z_2 = z_3$, but clearly $p(z_1) + p(z_2) < p(z_3)$. New cells can be inserted into the partition if new species are discovered (e.g., the species indicated by o_3).

Empty space from one point of view reflects the potential for adding new knowledge. From another point of view it can be seen as a matter of *hidden* knowledge. From this perspective it is as if we start from partitions which reflect the way God sees reality. Empty space then covers up what we humans do not yet know. The latter view was developed by Mislove et al. (1990) and resulted in the theory of partial sets. What we call empty space Mislove compares to packaging material. If we remove packaging material we potentially discover new things that were previously hidden from view. Our distinction between full and non-full partitions corresponds very closely to the distinction between 'clear' and 'murky' sets drawn in the theory of partial sets.

7.4 Fullness and emptiness

Since we have $Fullcell(z_1) \leftrightarrow (+_{z \sqsubset z_1} z) = z_1$, we also have $\neg Fullcell(z_1) \leftrightarrow (+_{z \sqsubset z_1} z) < z_1$. Consequently we can define what it means for x to be the empty space of the cell z_1 as follows. We first of all define x fills z_1:

DFills: $Fills(x, z_1) \equiv \neg Fullcell(z_1)$ and $x + (+_{z \sqsubset z_1} z) = z_1$

The empty space in z_1 is then its smallest filler and we define that x is the empty space in z_1 if and only if x fills the space not occupied by the subcells of z_1 and x is disjoint from all subcells of z_1:

DES: $ES(x, z_1) \equiv Fills(x, z_1)$ and $\forall y : Fills(y, z_1) \rightarrow x \leq y$

We note in passing that minimal cells, on the basis of the definitions above, are either empty (they do not project) or they are completely made up of empty space. ('Minimal' means: there is no further knowledge available, within a given partition, as concerns the objects onto which minimal cells are projected.)

ES determines the empty space of a cell uniquely. To prove this, assume $ES(x, z)$ and $ES(y, z)$. For simplicity we use 'c' as an abbreviation for '$(+_{z_1 \sqsubset z} z_1)$'. We can then rewrite $ES(x, z)$ as: $x + c = z$ and $\forall u : u + c = z \rightarrow x \leq u$, and rewrite $ES(y, z)$ as: $y + c = z$ and $\forall v : v + c = z \rightarrow y \leq v$. Since $z = u + c = v + c = x + c = y + c$ we have for all u, v: if $v + c = u + c = x + c = y + c$ then ($x \leq u$ and $y \leq v$). In particular this holds in the case of $\forall u, v : u = v$. Now assume that $x \neq y$. This means that for all u: if $u + c = x + c = y + c$ then ($x < u$ and $y \leq u$) or ($x \leq u$ and $y < u$). Without loss of generality consider for all u: if $u + c = x + c = y + c$ then ($x < u$ and $y \leq u$). Now without loss of generality consider for all u: if $u + c = x + c$ then $x < u$. Since for all u: $u + c = x + c$ holds by assumption it in particular this holds for $u = x$. Consequently we have $\exists u : u + c = x + c$ and $u = x$ and $u < x$ leading to a contradiction of the form $\exists u, x : u = x$ and $u \neq x$. Consequently we have $x = y$ as desired.

7.5 Cumulativeness

We can now define the notion of *cumulativeness*, which plays the same role in theory (B) which fullness plays in theory (A). The intuitive idea is as follows: a cell is cumulative relative to its immediate descendant cells if the mereological sum of the *projections* of these immediate descendants is identical to the projection of their partition-theoretic sum. For non-empty and non-minimal cells with at least two immediate descendants we define:

DCu1: $Cu(z_1) \equiv +_{z \sqsubset z_1} p(z) = p(\cup_{z \sqsubset z_1} z)$

One can see that $p(\cup_{z \sqsubset z_1} z) = p(z_1)$ holds under the given conditions. Consequently:

$$Cu(z_1) \leftrightarrow +_{z \sqsubset z_1} p(z) = p(z_1)$$

Again, $Cu(z_1)$ ensures that z_1 is cumulative relative to its immediate descendants. In order to ensure cumulativeness of a cell with respect to all its subcells, we define

DCu2: $Cu*(z_1, A) \equiv \forall z : z \subseteq z_1 \rightarrow (Cu(z) \text{ or } Min(z, A))$

A partition is cumulative if and only if all its cells are cumulative.

DCu3: $Cu(A) \equiv \forall z : Z(z, A) \rightarrow (Cu(z) \text{ or } Min(z, A))$

Another way of expressing this is as follows: a partition is cumulative if and only if it has a *basis in objects* (the objects projected by its minimal cells), and is then built up in stages in such a way that each non-minimal cell z projects onto the mereological sum of the objects projected by z's immediate descendants.

Recall that the notions of fullness and cumulativeness are intended to characterize partitions that have no redundancies of the sort defined in CR. A cumulative partition A is also exhaustive (CE_φ) with $\varphi = \exists z : Z(z, A)$ and $P(z, o)$.

7.6 Classes of partitions regarding fullness and cumulativeness

We can now distinguish four classes of partitions regarding their fullness and cumulativeness:

Full and cumulative. Consider a list of the 50 US States, divided into two sub-lists: the contiguous 48, the non-contiguous 2. Here the objects towards which this partition is directed are the States themselves under the obvious 'Utah'–Utah projection relation.

Full and non-cumulative. Consider Figure 7 (a). This partition represents the belief of
  some child who thinks that cats and dogs are the only animals there are.

Non-full and cumulative. Consider Figure 7 (b). This partition represents the way a
  child sees the world who does not understand the concepts of the Northern and
  Southern hemisphere and who thinks that there are places on Earth that are neither
  in the Northern, nor in the Southern Hemisphere, nor overlapping both, but are
  rather in some secret and wonderful land that has not yet been discovered.

Non-full and non-cumulative. Consider Figure 7 (c). Imagine that you have a terrible
  hangover, and your accounting of the people at the party last night consists of a
  root with three subcells: *you, John,* and *Mary.* You know that you are missing
  somebody, but you cannot remember who.

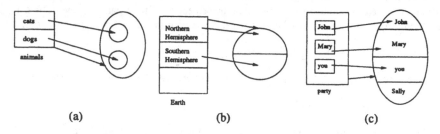

(a)   (b)   (c)

Figure 7: Examples of non-full and non-cumulative partitions

We humans are often aware of the fact that our partitions are not full or not cumulative,
or both. In the former case empty cells are often included into the cell structure in order to
make explicit our awareness of these shortcomings. Consider, for example, the periodic
table of the elements, which contains empty cells (labeled ununnilium, etc.), for elements
not yet synthesized. Moreover the fact that the periodic table in its current form has
a structure that can only accommodate a certain number of cells is not taken to imply
that in reality there could not be elements whose discovery would require the table to be
extended.

8   IDENTITY OF GRANULAR PARTITIONS

As a step towards a definition of identity for partitions Smith and Brogaard (2002b) propose a partial ordering relation between partitions, which they define as follows:

$$A \leq B \equiv \forall z : Z(z, A) \rightarrow Z(z, B)$$

They then define an equivalence relation on partitions as follows:

DE: $A \approx B \equiv A \leq B$ and $B \leq A$

Now, however, we can see that a definition along these lines will work only for partitions which are full. What, then, of those partitions which are equivalent in the sense of

DE but not full? What are the relationships between the presence of empty and redundant cells and the question of the identity of partitions? And what is the bearing on the question of identity of the phenomenon of empty space? Can partitions that have empty or redundant cells be identical? Can partitions which are not full be identical?

The question of whether or not partitions that have empty or redundant cells are identical cannot be answered without a theory of labeling. Corresponding empty cells in two distinct partitions, if they are to be considered as identical, need to have at least the same labels. We can only address this question informally here, leaving for another place the task of developing the necessary formal theory.

As already noted, one makes a different sort of error if one thinks that there are dodos from the error which one makes if one thinks that there is an intra-Mercurial planet. The two corresponding empty cells are thus distinct: they are signs of different sorts of failure arising when we project a partition onto reality.

Consider an inventory of the goods you plan to sell in your not-yet-established chain of beer-brewing stores. This is, surely, different from the inventory of the goods I plan to sell in my not-yet-established chain of wine-marketing stores. And this is so even in the event that our respective plans are never realized.

Consider the partition of the people in your building according to *number of days spent behind bars*. You can construct this partition—which is little more than a simple array of numbered boxes—prior to undertaking any actual inquiries as to who, among the people in your building, might be located in its various cells. Thus even before carrying out such inquiries you can know that this is a more refined partition than, for example, the partition of the same group of people according to *number of years spent behind bars*. The two partitions are distinct, and they will remain distinct even if it should turn out that none of the people in your building has spent any time at all in jail. In both cases all the people in your building would then be located in the cell labeled *zero* and all the other cells in both partitions would be empty. Yet the two partitions would be nonetheless distinct, not least because their respective maximal cells would have different labels.

We can now return briefly to our question whether partitions that are neither full nor cumulative can be said to be identical. One approach to providing an answer to this question would be to point out that, even though two partitions are outwardly identical, they might still be such that there are different ways to fill the corresponding empty space. Suppose we have what are outwardly the same biological taxonomies used by scientists in America and in Australia at some given time, both with the same arrays of empty cells. Suppose these partitions are used in different ways on the two continents, so that, in the course of time, their respective empty space gets filled in different ways. Were they still the *same* taxonomy at the start?

9   RELATED WORK

Variant forms of our granular partitions have been advanced elsewhere, in particular in the literature on Spatial Information Science. What makes spatial or geographic granular partitions particularly interesting is the fact they are very well structured in the sense that they not only obey our master conditions MA1–4 and MB1–6 but are also mereologically monotonic (CM). In addition they are often full and cumulative as well as exhaustive with respect to predicates such as land use, political affiliation, and so on.

Examples of geographic granular partitions are categorical coverages, which are thematic maps depicting the relationship of a property or attribute to a specific geographical area (Chrisman, 1982). A prototypical example of a categorical coverage is the land use map, in which a taxonomy of land use classes is determined (e.g., residential, commercial, industrial, transportation) and some specific area (for example a city) is then evaluated along the values of this taxonomy (Volta and Egenhofer, 1993). Another prototypical example is soil maps, which are based on a classification of soil covering (into *clay*, *silt*, *sand*, etc.). The zones of a categorical coverage are a jointly exhaustive and pair-wise disjoint subdivision of the relevant region of space (Beard, 1988). As discussed in (Bittner and Smith, 2001) categorical coverages can be understood in terms of the interplay of two granular partition: one targeting a region of space, the other targeting some attribute domain.

Another important feature of spatial granular partition is that they also recognize topological and geometric properties of the domain they project onto. In order to recognize structure beyond mereology the cell structure of a granular partition must have structural features in addition to the subcell relation between its cells. Examples of granular partitions that take topological structure, i.e. neighborhood relations between adjacent partition cells, into account can be found in Frank and Kuhn (1986); Bittner and Stell (1998); Erwig and Schneider (1997). Applications of granular partitions taking the ordering of the cell structure and the shape of cells into account were discussed in Frank (1992); Freksa (1992); Hernandez (1991).

Recently, spatial partitions were applied also to the representation of spatio-temporal phenomena (Erwig and Schneider, 1999) and temporal phenomena (Bittner, 2002). Independently, the notion of granularity was discussed in Stell (1999, 2000) and Stell, this volume.

10 CONCLUSIONS

This paper is a contribution to a formal theory of granular partitions. We defined master conditions that need to be satisfied by every partition. These master conditions fall into two groups: (A) master conditions characterizing partitions as systems of cells, and (B) master conditions describing partitions in their projective relation to reality.

At the level of theory (A) partitions are systems of cells that are partially ordered by the subcell relation. Such systems of cells are such that they can always be represented as trees, they have a unique maximal element and they do not have cycles in their graph-theoretic representations. But partitions are more than just systems of cells. They are also cognitive devices that are directed towards reality.

Theory (B) takes this latter feature into account by characterizing partitions in terms of the relations of projection and location. Cells in partitions are projected onto objects in reality. Objects are located at cells when projection succeeds. We then say that a partition *recognizes* the objects that are located at its cells. To talk of granular partitions is to draw attention to the fact that partitions are in every case selective; even when they recognize some objects, they will always trace over others.

Partitions do not only recognize objects, they are also capable of reflecting the mereological structure of the objects they recognize through a corresponding mereological structure on the side of their cell array. This does not mean, however, that all partitions actually do reflect the mereological structure of the objects they recognize. For it is an important feature of partitions that they are also capable of tracing over mereological structure. There are, for example, large classes of partitions that simply list objects, without caring at all how these objects hang together mereologically.

Our discussion of granularity showed that partitions have three ways of tracing over mereological structure: (1) by tracing over mereological relations between the objects which they recognize; (2) by tracing over the parts of such objects; (3) by tracing over the wholes which such objects form. The tracing over of parts is (unless mereological atomism is true) a feature manifested by every partition, for partitions are in every case *coarse grained*. The tracing over of wholes reflects the property of granular partitions of foregrounding selected objects of interest within the domain onto which they are projected and of leaving all other objects in the background where they fall in the domain of unconcern.

ACKNOWLEDGMENTS

Our thanks go to Berit Brogaard, Maureen Donnelly, Pierre Grenon, Jonathan Simon, and John Stell for helpful comments. Support from the American Philosophical Society, and from the NSF (Research Grant BCS-9975557: "Geographic Categories: An Ontological Investigation") is gratefully acknowledged. This work was also supported in part by DARPA under the Command Post of the Future Program, by the NSF Research on Learning and Education Program, and by the Wolfgang Paul Program of the Alexander von Humboldt Foundation.

REFERENCES

Beard, K. (1988). *Multiple representations from a detailed database: A scheme for automated generalization*. PhD thesis, University of Wisconsin, Madison.

Bittner, T. (2002). Approximate qualitative temporal reasoning. *Annals of Mathematics and Artificial Intelligence*. To appear.

Bittner, T. and Smith, B. (2001). A taxonomy of granular partitions. In Montello, D. R. (Ed), *Spatial Information Theory: Foundations of Geographic Information Science*, volume 2205 of *Lecture Notes in Computer Science*. Berlin: Springer-Verlag.

Bittner, T. and Stell, J. G. (1998). A boundary-sensitive approach to qualitative location. *Annals of Mathematics and Artificial Intelligence*, 24:93–114.

Casati, R. and Varzi, A. C. (1994). *Holes and other Superfacilities*. Cambridge, MA: MIT Press.

Casati, R. and Varzi, A. C. (1995). The structure of spatial location. *Philosophical Studies*, 82:205–239.

Chrisman, N. (1982). *Models of Spatial Analysis Based on Error in Categorical Maps*. PhD thesis, University of Bristol, England.

Degen, W., Heller, B., Herre, H., and Smith, B. (2001). GOL: A general ontological language. In Welty, C. and Smith, B. (Eds), *Formal Ontology and Information Systems*, pp. 34–46. New York: ACM Press.

Erwig, M. and Schneider, M. (1997). Partition and conquer. In Hirtle, S. C. and Frank, A. U. (Eds), *Spatial Information Theory: A Theoretical Basis for GIS*, volume 1329 of *Lecture Notes in Computer Science*. Berlin: Springer-Verlag.

Erwig, M. and Schneider, M. (1999). The honeycomb model of spatio-temporal partitions. In *International Workshop on Spatio-Temporal Database Management*, volume 1678 of *Lecture Notes in Computer Science*, pp. 39–59. Berlin: Spinger-Verlag.

Frank, A. (1992). Qualitative spatial reasoning about distances and directions in geographic space. *Journal of Visual Languages and Computing*, 3:343–371.

Frank, A. and Kuhn, W. (1986). Cell graphs: A provable correct method for the storage of geometry. In Marble, D. (Ed), *Second International Symposium on Spatial Data Handling*, pp. 411–436, Seattle, WA.

Frank, A., Volta, G., and McGranaghan, M. (1997). Formalization of families of categorical coverages. *International Journal of Geographic Information Science*, 11(3):214–231.

Freksa, C. (1992). Using orientation information for qualitative spatial reasoning. In Frank, A. U., Campari, I., and Formentini, U. (Eds), *Theories and Methods of Spatio-Temporal Reasoning in Geographic Space*, volume 639 of *Lecture Notes in Computer Science*. Berlin: Springer-Verlag.

Galton, A. (1999). The mereotopology of discrete space. In Freksa, C. and Mark, D. M. (Eds), *Spatial Information Theory: Cognitive and Computational Foundations of Geographic Science*, volume 1661 of *Lecture Notes in Computer Science*, pp. 251–266. Berlin: Springer-Verlag.

Hernandez, D. (1991). Relative representation of spatial knowledge: The 2D-case. In Mark, D. M. and Frank, A. U. (Eds), *Cognitive and Linguistic Aspects of Geographic Space*, pp. 373–386. Dordrecht, The Netherlands: Kluwer Academic Publishers.

Lewis, D. (1991). *Parts of Classes*. Oxford: Blackwell.

Mark, D. M. (1978). Topological properties of geographic surfaces: Applications in computer cartography. In *Harvard Papers on Geographic Information Systems*, volume 5. Cambridge, MA: Laboratory for Computer Graphics and Spatial Analysis, Harvard University.

Mislove, M. W., Moss, L., and Oles, F. J. (1990). Partial sets. In Cooper, R., Mukai, K., and Perry, J. (Eds), *Situation Theory and Its Applications I*, volume 22 of *CSLI Lecture Notes*, pp. 117–131. Stanford, CA: Center for the Study of Language and Information.

Schuhmann, K. and Smith, B. (1987). Questions: An essay in Daubertian phenomenology. *Philosophy and Phenomenological Research*, 47:353–384.

Searle, J. R. (1983). *Intentionality. An Essay in the Philosophy of Mind*. Cambridge: Cambridge University Press.

Smith, B. (1991). Relevance, relatedness and restricted set theory. In Schurz, G. and Dorn, G. J. W. (Eds), *Advances in Scientific Philosophy. Essays in Honour of Paul Weingartner*, pp. 45–56. Amsterdam/Atlanta: Rodopi.

Smith, B. (1998). The basic tools of formal ontology. In Guarino, N. (Ed), *Formal Ontology in Information Systems*, Frontiers in Artificial Intelligence and Applications, pp. 19–28. Amsterdam: IOS Press.

Smith, B. (1999). Truthmaker realism. *Australasian Journal of Philosophy*, 77(3):274–291.

Smith, B. (2001a). Fiat objects. *Topoi*, 20(2):131–148.

Smith, B. (2001b). True grid. In Montello, D. R. (Ed), *Spatial Information Theory: Foundations of Geographic Information Science*, volume 2205 of *Lecture Notes in Computer Science*, pp. 14–27. Berlin: Springer-Verlag.

Smith, B. and Brogaard, B. (2002a). Quantum mereotopology. *Annals of Mathematics and Artificial Intelligence*, 34:153–175.

Smith, B. and Brogaard, B. (2002b). A unified theory of truth and reference. *Logique et Analyse*. To appear.

Stell, J. G. (1999). Granulation for graphs. In Freksa, C. and Mark, D. M. (Eds), *Spatial Information Theory. Cognitive and Computational Foundations of Geographic Information Science*, number 1661 in Lecture Notes in Computer Science, pp. 417–432. Berlin: Springer-Verlag.

Stell, J. G. (2000). The representation of discrete multi-resolution spatial knowledge. In *Proceedings of Seventh International Conference on Principles of Knowledge Representation and Reasoning (KR2000)*, San Francisco. Morgan Kaufmann Publishers.

Stell, J. G. (2002). Granularity in change over time. In Duckham, M., Goodchild, M. F., and Worboys, M. F. (Eds), *Foundations in Geographic Information Science*, pp. 95–115. London: Taylor & Francis.

Volta, G. and Egenhofer, M. (1993). Interaction with GIS attribute data based on categorical coverages. In Frank, A. U. and Campari, I. (Eds), *Spatial Information Theory: A Theoretical Basis for GIS*, volume 716 of *Lecture Notes in Computer Science*. Berlin: Springer-Verlag.

Wittgenstein, L. (1961). *Tractatus Logico-Philosophicus*. London: Routledge and Kegan Paul. Translated by McGuinness, M. F. and Pears, D.

CHAPTER 8

On the Ontological Status of Geographical Boundaries

Antony Galton
School of Engineering and Computer Science
University of Exeter, Exeter, EX4 4QF, UK

1 INTRODUCTION

Boundaries occupy a curiously ambivalent position in any geographical ontology. On the one hand, it seems uncontentious that the primary spatial elements of geography are regions of various kinds: regions are where we live and where things are located. From this point of view, boundaries are only of interest because they define the limits of regions. But precisely because of this, boundaries can acquire a life of their own. The existence of a boundary can have a palpable effect on the behaviour of objects and people in its vicinity. Disputes over territory automatically become focussed into disputes over boundaries, and the boundary itself can become a symbol for the territory it delineates: 'Not only do boundaries give the country a shape, but they suggest a uniformity within that shape which separates it from the outside, from what is alien and foreign' (Dorling and Fairbairn, 1997). Indeed, in ordinary speech there is a slippage between 'within this region/area/territory' and 'within these boundaries/limits/borders', pointing to the ease with which we can pass between thinking in terms of regions and thinking in terms of boundaries. The history of language itself can illustrate this. The English word *town*, for example, is derived from an Old English word *tun*, meaning an enclosure. It is related to the Dutch *tuin* which means a garden, an enclosure containing trees, grass and flowers rather than streets and buildings. The original meaning of the word appears to have been not the area enclosed but the fence or hedge which does the enclosing. This meaning persists in the cognate German form *Zaun*, which refers to the fence or hedge itself, not an enclosed area.

Boundaries embody many different functions. A boundary may be erected in order to keep captives in or to keep intruders out, or simply to prevent mixing. But boundaries can be crossed: physical boundaries such as walls usually include gateways or portals by which movement across the boundary is simultaneously facilitated and regulated. Thus there is another slippage in our thinking, between borders and border-crossings: 'we'll reach the border soon' is said in the expectation of crossing it. Consider too how the notion of a *screen* has passed from being a barrier (as in a fire screen) to a bearer of images

151

(painted on the fire-screen, or projected onto canvas) and thence to a kind of window into another world (television and computer screens).

The ambivalence of boundaries extends to their representation, both in maps and in information systems. Boundaries of a sort may be present in a representation even when the representation has no explicit symbols for them. As a child I had a wooden jigsaw puzzle, with pieces shaped like the counties of England and Wales. Each piece represented a county; so one could say that the *edge* of the piece represented the boundary of the county. But boundaries are not just 'boundaries *of*', they can be 'boundaries *between*' as well. When the puzzle was assembled, the boundaries between the counties were represented, implicitly, by the interstices between neighbouring pieces. So the puzzle showed the boundaries even though it contained no explicit boundary symbols. Likewise, in a map, the boundary between neighbouring regions can be shown implicitly as the line of contact between two differently coloured areas, or explicitly as a printed line. In vector-based geographical information systems, it is usual to specify the location of a region by specifying the location of its boundary. Does this mean that such systems necessarily include boundaries as elements within their ontology?

In this chapter I address some of the problems posed by geographical boundaries. I begin with a survey of different kinds of boundaries in an attempt to develop a reasonably comprehensive classification. I then consider relationships between the different kinds of boundary, examining how boundaries of one type can evolve into or otherwise give rise to boundaries of other types. This is followed by a discussion of some of the key properties of boundaries that are of relevance to the problems of how they should be represented. Finally, I examine how boundaries can be represented within the two main paradigms of geographical information science, the field-based and the object-based.

2   CLASSIFICATION OF BOUNDARIES

In this section I shall attempt a broad-brush classification of boundaries. The classification takes the form of a hierarchical tree structure, as shown in Figure 1. The top-level distinction is between *physical* and *institutional* boundaries. This corresponds closely with the distinction between 'bona fide' and 'fiat' boundaries introduced by Smith (1995), although his distinction and mine do not agree in every respect. Physical boundaries are in turn divided into *material* and *epiphenomenal* boundaries. In the case of the former, there is some material substance or phenomenon which constitutes the boundary, and the location of the boundary is the location of its material or phenomenal constituents. An epiphenomenal boundary depends on matter for its existence but has no material or phenomenal substance in itself. Each of these subclasses is itself subdivided; the details are given below. As is usual with the classification of a rich and varied domain, the distinctions here drawn are not in every case entirely clear-cut: some cases can be classified in different ways depending on how they are interpreted, and we may find intermediate cases which seem to occupy a middle ground between two positions in the classification.

2.1   Physical boundaries

All boundaries exist by virtue of the distribution of matter and energy in space and time, but boundaries may differ as to just how their existence depends on such distribution. For

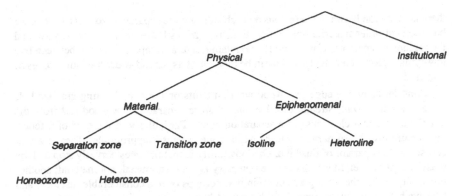

Figure 1: Classification of boundaries

some boundaries, for example the International Date Line, the dependence of the boundary on the material facts is mediated by individual or collective human intentionality. These are institutional boundaries; all other boundaries are physical boundaries. Since the latter class is defined by exclusion, i.e., in terms of what they are not, it will turn out to be rather heterogeneous.

Amongst physical boundaries we distinguish *material* boundaries, which are, so to speak, *made of matter*, from *epiphenomenal* boundaries, which exist by virtue of the distribution of matter in space and time but are not themselves made of matter. Consider an area of woodland separated from the adjacent grassland by a wall. The wall is a material boundary between the woodland and the grassland. If it is removed, we may still speak of the boundary between the two regions, but now there is nothing material we can point to and say 'this is the boundary'. The boundary now exists solely by virtue of the distribution of woodland and grassland in that area: it is epiphenomenal. Similarly, the peel of an orange is a material boundary between the interior of the orange and the outside world; the surface of the orange (which is also the surface of the peel) is an epiphenomenal boundary between the orange and the outside world.

2.1.1 Material boundaries

We may distinguish two kinds of material boundary, which I call *separation zones* and *transition zones*. In both kinds, the boundary occupies a zone—usually a ribbon-like band much longer than wide—whose material or phenomenal contents differ in character from those of the regions on either side. The distinction between the two kinds hinges on the nature of this difference in character. In a transition zone, the character is intermediate between that of one side and that of the other. In many cases there is a smooth gradation in character from one side, across the boundary, to the other. There are many different kinds of transition zone, according to the profile of the gradation; a preliminary classification is given by Plewe (1997). In a separation zone, the character of the zone is distinct from, and not intermediate between, the characters of the regions it separates. The separation zone may often be thought of as a *barrier*; but this is primarily a *functional* notion, characterised in terms of affordances rather than material constitution. If there is gradation between the regions on each side and the separation zone itself, then we have second-order material boundaries, transition zones between a separation zone and the regions it separates. If we regard a mountain range as a separation zone between two low-lying regions, then the foothills can be thought of as second-order transition zones of this kind.

Consider again the edge of a wood where it abuts on the neighbouring grassland. If there is a fence, wall, stream, or other such feature separating the woodland from the grassland, this is a clear case of a separation zone. Particularly in the case of a fence, this feature may be so narrow that the word 'zone' seems inappropriate. There may, of course, be no separator of this kind. One possibility is that the trees abruptly stop and the grassland takes over, in which case the boundary is epiphenomenal, not material; another possibility is that the trees gradually thin out, perhaps over a considerable distance, and here we have a transition zone. The distinction between separation zones and transition zones is not always clear-cut. Suppose there is a scrubby area populated by bushes between the woodland proper and the grassland. Is this scrub zone intermediate in character between the woodland and the grassland or not? No doubt many different aspects could be considered here, some of which would lead one to conclude that the scrub zone is a transition zone, others that it is a separation zone.

We can further subdivide separation zones according to the nature of what is separated. There seems to be a considerable difference between a river meandering through an otherwise uniform plain and a sandy beach separating sea and land. Both can function as boundaries, indeed as barriers, but whereas the river separates regions which are not in themselves distinct in character, the beach separates regions which could hardly be more different. Both are separation zones (though the beach also has a certain transitional character, as discussed more fully below), yet they do not seem to belong together. I therefore classify separation zones into *homeozones*, which separate like from like, and *heterozones*, which separate unlike regions. Since there can be degrees of likeness, there is a gradation between these two types; none the less, there are many clear-cut cases. The protective outer coverings of many everyday objects—the peel of an orange, the walls of a house, the bark of a tree, human skin—are clear cases of heterozones. In many cases features such as roads, railways, rivers, fences, and hedges are homeozones, although of course any of these can also be a heterozone, it being not intrinsic to their nature that they should separate either like from like or unlike from unlike. The walls of a house illustrate this: the exterior walls are heterozones, separating 'indoors' from 'outdoors', but the interior walls are homeozones, separating room from room. But if we were to focus on the different characters of different rooms, then we could come to see these internal walls as heterozones also.

2.1.2 *Epiphenomenal boundaries*

One kind of epiphenomenal boundary is an *isoline* for a field, defined as 'the locus of all points in the field with the same attribute value' (Worboys, 1995). This need not, of course, be a line, since there could be an area throughout which the field has constant value. Familiar examples of isolines are contours (isolines of elevation), isotherms, and isobars. In themselves they do not partake of the character of boundaries, except insofar as *any* line can be regarded as a boundary; but in particular cases they can give rise to more overtly boundary-like phenomena; some examples will be given below.

Epiphenomenal boundaries can also be defined in terms of what they separate rather than what they join, and in this case they are closer to our normal understanding of a boundary. An example is the 'isogloss' of dialectology, which is a line dividing areas with different speech forms. Despite their name, these are not isolines in the accepted sense, and for this reason some dialectologists use the more satisfactory term 'heterogloss'. In keeping with this, I shall use the term *heteroline* to refer to any line of separation between areas of different attribute values. The attribute takes some constant value X on one side of the line and a different constant value Y on the other side. All lines of latitude are isolines with respect to the continuous variable 'latitude', but certain geographically salient lines of latitude such as the equator, the tropics of Cancer and Capricorn, and the arctic and antarctic circles may be regarded as heterolines with respect to qualitative features of the annual apparent motion of the sun. Lines of longitude are more artificial, involving as they do an arbitrarily chosen base line, but they too can be regarded as examples of isolines, albeit rather marginal.

In all these cases the boundary is real but lacks physical substance; it is located in space but does not occupy space. It arises as a by-product of particular distributions of matter or energy (including human behaviour) over space and time. It should be emphasised that these are *not* 'fiat' boundaries in the sense of Smith (1995): although the equator is always described as an 'imaginary' line, it has a physical reality (even though no physical substance) that is independent of any human cognitive acts (although in common with many more substantial things it may require specific types of cognitive activity to *discover* it or think it worth considering). Still less are they the product of human intentions.

Epiphenomenal boundaries can exist as a result of human behaviour. Linguistic boundaries are of this kind, both boundaries between sharply distinct languages (e.g., between French and Flemish in Belgium), and dialect boundaries within a single language or closely related group of languages (e.g., between Low German and High German). These boundaries arise as an epiphenomenon of human behaviour patterns, but epiphenomenal boundaries can also *affect* human behaviour. Jones (1945) cites an interesting example given by F. Kingdon Ward, who noted that for the Tibetans, the 'invisible' barriers of the 50-inch rainfall contour and the 75 per cent saturated atmosphere are 'far more formidable ... than the Great Himalayan range'.

The 'visibility' or otherwise of a boundary is, of course, relative to a means of 'seeing': implicitly, normal human perception. Imagine a piece of paper, so coloured that, when illuminated in white light, the left half reflects only monochromatic yellow light of wavelength 580nm, and the right half reflects a certain mixture consisting of red and green lights, of wavelengths 670nm and 540nm respectively. The proportion of the mixture can be selected so that the two sides of the paper are indistinguishable to humans with normal colour vision (Hurvich, 1982). Such a person would fail to see the boundary between the two halves of the paper, but that boundary certainly exists in physical reality. It can *become* visible, however: if the paper were viewed through a filter which absorbed light of wavelengths longer than 600nm, then while the left half would still appear yellow, the right half would now appear green, with the boundary between the two clearly visible. Thus the presence or absence of a physical boundary is a separate matter from its visibility or invisibility to a given observer under given conditions.

Often, visible epiphenomenal boundaries are caused by invisible ones. A common case is an isoline (or isoline bundle) giving rise to a heteroline. The area inhabitable by a given plant species will be limited by epiphenomenal temperature boundaries (isotherms), amongst other factors. The tree line encountered as one ascends a mountain, or moves north towards arctic regions, while still epiphenomenal, is the visible expression of a bundle of thermal boundaries pertaining to a variety of tree species. As usual with such boundaries, the details are complex and poorly understood (Tivy, 1971). Tree lines may be more or less sharp: in Europe, the altitudinal tree-line is much more sharply defined in the Alps than in the maritime north-west (Collinson, 1988). Likewise, altitudinal treelines are in general sharper than latitudinal ones. In the northern Alps, 'only a very narrow zone consisting of stunted, low forms provides the transition from forest to treeless alpine belt' (Walter, 1985, p.209), whereas the transition from boreal forest to treeless tundra 'may extend for hundreds of kilometres' (ibid., p.283). The existence of such a transition zone suggests the possibility of defining *two* lines, one to mark the boundary between the fully-fledged forest and the transition zone, and one to mark the boundary between the latter and the treeless area. Pears (1977) uses the terms 'timberline' and 'treeline', often treated as synonymous, to distinguish these, modelling his terminology on that used by some European ecologists: *Waldgrenze* ('the upper limit of tall, erect tree growth occurring at forest densities') and *Baumgrenze* ('the line through the last scattered trees on the mountain slopes'). The zone between the two lines is known as the *Kampfzone*.

Something similar occurred with Wallace's line, the boundary originally drawn by Alfred Russel Wallace in Indonesia to separate the area where the fauna is predominantly Asian (Oriental) in character from the area where it is Australian. He found the line to be particularly sharp between the islands of Bali and Lombok, only twenty miles apart but with very distinct faunas. North of there, Wallace drew the dividing line between Borneo and Sulawesi, and between the Philippines and the Moluccas. However, later observations led various researchers to draw the line in different positions; one such line, Weber's line, ran further east than Wallace's line; between them the faunas display intermediate character between the Oriental fauna to the west of Wallace's line and the Australian fauna east of Weber's line. Pielou (1979) declares that 'the whole zone, rather than either of the lines, is the true boundary separating the Oriental and Australasian regions'; the name Wallacea is sometimes given to this zone (Putman, 1984). Since this 'true boundary' is in fact an area, it should, like the *Kampfzone* of the forest margin, be regarded as a material boundary, not an epiphenomenal one. Wallacea is thus a region which can be characterised by the fact that its fauna involves an admixture of Oriental and Australasian elements, just as the *Kampfzone* is characterised by the fact that although it is inhabited by trees, they do not grow at forest densities, and many of them are 'deformed, twisted, knarled, dwarfed, or prostrate in appearance due to the severe environmental regime at these levels' (Pears, 1977).

A series of epiphenomenal boundaries running roughly parallel to one another gives rise to a zonation. The sea-shore is a good example of this. The epiphenomenal boundaries in question are determined by the variation in tidal levels. Yonge (1949) identifies the shore as the region bounded by the extreme high water level of spring tides (EHWS) and the extreme low water level of spring tides (ELWS), and uses the average high tide level (AHTL) and the average low tide level (ALTL) to divide this into three zones, called the upper shore, middle shore, and lower shore, stating that '[t]his subdivision of the shore does appear best to correspond to our present knowledge of the vertical distribution of the shore population, while it has the merit of much greater simplicity than many previous schemes of subdivision'. Lewis (1964), however, does not find tidal zonation to be an adequate basis for characterising shore ecology, rather taking the view that 'the zones are biological entities which can only be defined by biological means' (Lewis, 1964, p.48). Thus Lewis's key boundaries are the upper limit of the *Littorina/Verrucaria* belt, the upper limit of barnacles, and the upper limit of *Laminaria* (i.e., kelp), these three boundaries defining a 'littoral zone' divided into the 'littoral fringe' and the 'eulittoral zone'. The position of these boundaries in relation to the EHWS and ELWS tidal levels varies greatly from shore to shore, being particularly sensitive to how sheltered or exposed the shore is.

Epiphenomenal boundaries exist in infinite abundance: we have only to define them. This reflects the continuous variation of so many of the measurable values in nature. If it suited us, we could define any number of other tidal levels on the shore, for example the points uncovered by 20 per cent of tides. We can choose any temperature we like and define the isotherm for that temperature; we can pick out the contour for any elevation. The boundaries we choose to define are motivated in some way, for example, because they are easy to define, using simple concepts like mean or extreme values, or 'round' numbers; or because they correlate with some visible features of the environment. Note that the virtue Yonge claims for his chosen subdivision of the shore comprises both these elements: it is simple, and it correlates with ecological zonation. To the extent that motivation by simplicity dominates, the epiphenomenal boundary acquires something of a 'fiat' character; to the extent that motivation by correlation with nature dominates, this fiat character may be negated. Thus epiphenomenal boundaries in general may be thought to occupy an ambiguous intermediate position between the most bona fide physical boundaries, and purely institutional ones.

2.2 Institutional boundaries

Institutional boundaries are those which are stipulated to exist by human fiat, for example in accordance with the terms of a peace treaty. They include all international boundaries and also intranational boundaries such as those between administrative regions, and those defining land ownership. Even where such a boundary is stipulated to follow some pre-existing physical boundary, it must still count as institutional. It is by human fiat, for example, that the boundary between the English counties of Devon and Cornwall is stipulated to follow the river Tamar. The fiat which legitimises the boundary also stipulates the coincidence of the fiat boundary with the pre-existing bona fida boundary. In fact, the strength of the attachment can be quite precisely defined in law: 'natural and gradual shifts of course move the boundary with the stream; changes in a river because of artificial alterations, or owing to sudden floods, do not' (Harley, 1975).

The terms 'accretion' and 'avulsion' have been used to refer respectively to gradual and sudden changes in the course of a river. The latter would include, in addition to artificial alterations and sudden floods, the cutting off of a meander to form an ox-bow lake, a phenomenon that occurred with some regularity in the Rio Grande along the USA-

Mexico boundary, leading to the formation of pieces of land ('bancos') which belonged to the state on the other side of the river until the boundary was restored to the actual river course in 1905. Subsequent engineering works simplified the course of the river, leading to further boundary adjustments (Boggs, 1940; Jones, 1945; Prescott, 1965).

Institutional boundaries are generally conceptualised as lines in the Euclidean sense, i.e., as having length but no breadth. If a boundary is truly a line then the precision with which its location can be ascertained is limited only by our measurement capacities. Contrast this with the case of, say, a tree line, which it does not make sense to locate more precisely than the width of a typical tree, i.e., well within our measuring abilities—so that ascertaining the location of a tree line is not a problem of measurement but of definition.

There are some situations in which we seem to need to determine boundaries to arbitrarily fine precision. Ball games such as tennis are a case in point. Whether a ball is 'in' or 'out' depends on which side of a boundary line it first strikes the ground. The difficulty of ascertaining this in critical cases leads to frequent disputes. For the individual players, a good deal can hang on the outcome of such disputes: the loss of a game, a set, a match, a tournament, and all that that entails in terms of financial reward and public prestige. Tennis matches proceed under the implicit assumption that any ball is determinately either in or out, and it is up to us to determine which of these cases holds. When the ball lands close to the boundary, it becomes very unreliable to judge this by eye, and hence it is customary to employ automatic sensors. The reliability of these sensors is often questioned, not least by the players themselves.

The nature of fiat boundaries may be clarified by reference to Searle's theory of *institutional facts* (Searle, 1995). Searle lists six properties which characterise institutional facts, namely:

1. The self-referentiality of many social concepts (part of what makes an institutional fact true is the fact that some social group holds it to be true).
2. The use of performative utterances in the creation of institutional facts (e.g., 'I appoint you chairman', 'War is hereby declared').
3. The logical priority of brute facts over institutional facts.
4. Systematic relationships amongst institutional facts.
5. The primacy of social acts over social objects, of processes over products.
6. The linguistic component of many institutional facts.

With reference to international boundaries, we may note that (1) if *all* social groups cease to believe in the existence of some boundary, then that boundary no longer exists, even if the associated physical paraphernalia (fences, border posts, and the like) persist; (2) many international boundaries are brought into existence by the signing of a bilateral agreement between the parties concerned, and these signings function as performative utterances; (3) underlying their performative character, however, is the brute fact that on one occasion certain humans made particular marks on paper, and later, when the boundary was demarcated on the ground, various erections of stone or barbed wire, etc., were constructed along a linear series of spatial locations; (4) the existence of an international boundary is bound up with an intricate network of trade agreements, immigration procedures, social relations, etc., which together constitute a system; (5) the boundary, as an object, may, to paraphrase Searle, be regarded as in a sense just the 'continuous possibility' of the activities characteristically associated with boundaries, such as the formalised boundary-crossing procedures and the deflection of trajectories that would otherwise cross over the line delineated as the boundary; and finally, (6) the definition of the boundary as expressed in the signed agreement is partly constitutive of the fact of the boundary's existence.

An institutional boundary is distinct from any correlated physical boundaries. Searle imagines a 'primitive tribe' that builds a wall around its territory. This wall functions as a boundary 'in virtue of sheer physics': so far, it is not an institutional boundary. He now supposes that the wall 'gradually evolves from being a physical barrier to being a symbolic barrier', decaying to leave only a line of stones. If the inhabitants and their neighbours continue to *recognize* the line of stones as marking the boundary of the territory, it then functions as a boundary not in virtue of physics but 'in virtue of collective intentionality' (Searle, 1995, p.39f). Searle contrasts the case in which the members of the tribe 'simply have a disposition to behave in certain ways', so that their behaviour is just like that of animals marking the limits of their territory, with the case in which they 'recognize that the line of stones creates rights and obligations, that they are *forbidden* to cross the line, that they are *not supposed* to cross it', in which case the stones now 'symbolize something beyond themselves; they function like words' (ibid., p.71). In the former case we have an epiphenomenal rather than an institutional boundary. Searle's remarks indicate that there need be no clear dividing line between epiphenomenal boundaries and institutional ones.

3   DEPENDENCIES AMONGST BOUNDARY TYPES

I have referred several times to cases in which a boundary of one type can evolve into or otherwise give rise to a boundary of another type. Here I survey the range of possibilities a little more systematically. Since my classification recognises six different types of boundary, there are thirty possible types of transition of the form 'boundary of type X gives rise to boundary of type Y'. I do not discuss all thirty types separately: many of them can be considered together.

3.1   Institutional boundaries arising from physical boundaries

Searle's example of a physical wall decaying to a line of stones which is still respected as a boundary, having deontic rather than physical force, is a clear case of this category of transition. More generally, there are many examples where there appear to be 'natural' lines along which to draw an institutional boundary.

The coastline forms a natural boundary for any nation whose territory abuts the sea. For an island nation, this may form its entire boundary. Even in this case, though, we cannot simply identify the nation's boundary with its sea coast. The boundary of the nation (as opposed to that of the island) is institutional. It may be stipulated to follow the coast, but inevitably there are complications. International law distinguishes inland waters, territorial waters, and the high seas, and disputes often occur as to the exact locations of the boundaries between them. For example, it is usual to draw the boundary between territorial waters and high seas three nautical miles out from the low-water mark of the shore. Inland waters include so-called 'true bays'; but the definition of a true bay is contentious. In 1930, the Hague Conference for the Codification of International Law ruled that an indentation between two headlands less than 10 miles apart was to be classified as a 'closed bay' if its area exceeded that of the semicircle with the distance between the headlands as diameter, and an 'open bay' otherwise. With closed bays, the seaward boundary of the inland waters is taken to be the straight line joining the headlands, whereas with open bays that boundary is defined as the low-water mark following the sinuosities of the coast (Shalowitz, 1962).

Mountain ranges provide another 'obvious' type of physical boundary along which to draw an institutional boundary, but here too there are complications. Should the boundary follow the crest formed by the highest peaks, or the associated water-parting? In general these lines do not coincide; indeed, they may be many miles apart. Choosing one rather than the other can have curious consequences. In the Pyrenees, the boundary between France and Spain was settled by a treaty of 1659 to follow the line of the high peaks. This placed the town of Llívia, in the upper valley of the Spanish Ro Segre, on the French side of the border, even though in its economic and political affiliations it was Spanish rather than French. To rectify this a boundary was drawn round Llívia to form a Spanish enclave entirely surrounded by French territory, a status it has retained to this day—as documented at http://www.llivia.com/ (Peattie, 1944).

River boundaries are equally problematic, as we have already seen in relation to the Rio Grande. Once it has been decided that an institutional boundary is to follow a river, there remains the choice of exactly what line along the river it is to follow. There are three 'natural' choices: one of the banks of the river, the 'median line' (i.e., the locus of points equidistant from the two banks), and the 'thalweg' (i.e., the locus of lowest points of successive cross-sections along the river). The bank, if precisely defined, is a heteroline, otherwise a heterozone or perhaps a transition zone; the median line and thalweg are isolines (although these are not always well-defined). A bank is the most straightforward to identify *in situ*, but it leads to a gross asymmetry: one of the countries effectively owns the river. This is the case with the French-German border where it follows the course of the Rhine: the boundary is stipulated to follow the west bank, so the river itself is German.

Another frequent choice of line for an institutional boundary to follow are the astronomically defined 'imaginary' lines on the earth's surface: lines of latitude and longitude. These are isolines. Boundaries of this kind are frequent amongst the American state boundaries, as well as providing a substantial part of the international boundary between Canada and the USA, which in its western portions follows the 49°N line of latitude—even to the extent of cutting off the tip of a peninsula extending southwards from Canadian territory, giving rise to a small US enclave, Point-Roberts.

In general, institutional boundaries are defined in terms of epiphenomenal rather than material physical boundaries. An institutional boundary should ideally take the form of a line rather than a zone, for a boundary zone must be declared neutral, and as such is susceptible to conflicting territorial claims from either side. It is pertinent here to recall the four main stages of international boundary-making, as expounded by Jones (1945): first, political decisions on the allocation of territory; second, delimitation of the boundary in a treaty; third, demarcation of the boundary on the ground; and finally, administration of the boundary. At the first stage it is not unusual to allocate territory in accordance with some natural barrier such as a river or mountain range, but at the second stage greater precision is required. This is where the decision is made whether to follow the line of high peaks or the water-parting, or with a river boundary the median line, the thalweg, or one of the banks. There is a natural progression from a material boundary to an epiphenomenal one; such progressions form the topic of the next section.

3.2 Physical boundaries arising from other physical boundaries

3.2.1 *Epiphenomenal boundaries arising from material boundaries*

Epiphenomenal boundaries must arise from material circumstances of some sort, and these 'material circumstances' often amount to the existence of a material boundary. Any reasonably sharply defined separation zone, say between regions A and B, gives rise to two heterolines, representing the boundary between the separation zone and A, and the boundary between the separation zone and B. The banks of a river, where these are sharply defined, are of this kind, likewise the edges of a road and the two sides of a wall. If the edges of the separation zone are graded then they should be classified as secondary transition zones; this would be the case where a river does not have well defined banks but is separated from the dry land by a marshy area of intermediate character.

A transition zone itself gives rise to epiphenomenal boundaries. If the transition is smoothly graded then there are of course innumerable isolines. But the transition zone as a whole may be supposed to have reasonably clear-cut edges, since it represents the transition between two regions each with its own well-defined character. The edges of the transition zone are where this character first begins to change in the direction of the character of the region on the other side of the transition. This may be a purely statistical effect: where the speakers of a language A meet the speakers of language B, there might be a transition zone within which speakers of the two languages intermingle. As we move across the zone from the A side to the B side, the percentage of B-speakers increases (not necessarily smoothly) from 0% to 100%. Thus defined, the edges of the zone are isolines, but they may also be regarded as heterolines.

3.2.2 *Heterolines arising from isolines*

In fact, any isoline can be regarded as a heteroline. The $x = k$ isoline (where x is a continuous field variable and k is one of its values) is the heteroline separating areas having the character $x < k$ from areas having the character $x > k$. This becomes significant if these two numerically defined characters correspond to or give rise to something more qualitatively salient. In this way, a sharp treeline, where that exists, is a heteroline determined by isotherms.

Even where such qualitative significance is lacking, it is sometimes useful to partition the range of a continuous field variable into a discrete set of qualitative bands, replacing an infinite number of isolines by a finite number of heterolines giving a partial representation of the field. One could regard the elevation contours in a map as being either a selection of isolines (corresponding to, say, 50m, 100m, 150m, etc.), or a set of heterolines (separating 'lower than 50m' from 'higher than 50m', 'lower than 100m' from 'higher than 10m', etc.). In some maps this is made more vivid by assigning different colours to the zones thereby defined. Likewise the Arctic Circle is both an isoline, the locus of points having latitude 66.53°N, and a heteroline, separating those places where the sun is never visible at midnight from those where it sometimes is.

3.3 Physical boundaries arising from institutional boundaries

One way in which an institutional boundary can give rise to a physical boundary is through the demarcation of the former, for example the construction of a fence, wall, or other barrier. As Smith (1995) puts it, 'boundary-markers ... tend in cumulation to convert what is initially a fiat boundary into something more real'. More subtle are the long term effects that may arise from the simple existence of the boundary, regardless of how it is physically demarcated. An international boundary drawn across an initially homogeneous region may lead in time to the two subregions thereby separated acquiring marked differences in character. Prescott (1965) cites a detailed study by Daveau of the effects of the Swiss-French boundary in the Jura to the west of Lake Neuchâtel over several centuries, where, for example, the 'small strip fields' of Amont in France are clearly contrasted with the 'summer pastures' of Carroz in Switzerland, 'even though the physical character of the landscape on both sides is the same' (Prescott, 1965, p.97).

This is a case where the physical effect of a boundary is something directly visible, but in many cases the effect is more subtle, having to do with patterns of movement and interaction. A political boundary introduces discontinuities into such patterns, some of which may be obvious, others only to be uncovered by detailed research. Fielding (1974), for example, notes that '[p]eople in Vancouver ... are more likely to marry partners from Toronto or Winnipeg than from Seattle, Tacoma, or Bellingham, despite the proximity of single people in the latter group.' Such differences in patterns of interaction are epiphenomenal. One could, albeit rather artificially, consider a field whose value at a given point is the probability that a randomly-chosen single inhabitant of Vancouver will marry a person living at that point. Then the state of affairs described by Fielding would show up as a heteroline marking a discontinuity in the field values along the border.

4 SOME KEY ATTRIBUTES OF BOUNDARIES

In the light of the foregoing discussion it should be evident that the world of geographical boundaries is highly diverse, encompassing physical, biological, psychological, social, and political phenomena. Yet if we are to represent boundaries in an information system we must extract from this diversity some more general principles that apply across a wide range of cases. The classification introduced earlier is a first step in this direction; I now turn to an examination of some key features of boundaries that a model should take into account.

4.1 Dimension

Modelling geographical entities usually begins with a top-level classification of such entities according to dimension: there are point objects (that is, objects conceptualised as points), line objects such as roads, rivers, railways, coastlines, contours, and boundaries, and areal objects ('regions'). In such a classification, geographical boundaries are naturally assigned to the category of line objects. This category is, however, rather diverse: Mark and Csillag (1989) list five distinct types of 'geographic lines', two of which are further subdivided into two. Boundaries occupy two of these types, which they designate 'legislated line' (including 'some political boundaries') and 'area-class boundary' (e.g., climatic, vegetation, and soil-type boundaries).

Geographic lines represent entities which, at least after idealisation, are intrinsically one-dimensional, that is, positions along the line can be uniquely specified by means of a single numerical variable even though the line itself may weave a complex path in two or three dimensions. The notion of idealisation is important here: many 'lines' are in reality areas or volumes. They can be idealised as lines because all but one of their intrinsic dimensions are of negligible size in comparison with the remaining one: a river, for example, typically has a length of tens, hundreds, or even thousands of kilometres, whereas its width is of the order of one kilometre or less, and its depth of the order of tens of metres at most.

Any geographic line can be *thought of* as a boundary: it is the boundary between the area on one side and the area on the other. Whether or not it *functions* as a boundary depends on a variety of factors. As a first high-level generalisation, a line can be conceived in two ways: from the point of view of possible motion *along* it, and from the point of view of possible motion *across* it. Conceived in the first way, a line is a *way* or *path*; in the second, a *boundary*, *barrier* or *gateway*. As Couclelis and Gottsegen (1997) put it, 'a freeway is a *way* or a *barrier* depending on which way you look'. Many boundary functions are therefore defined in terms of 'across' rather than 'along'. They have to do with how a boundary regulates movement or communication across it. Examples of such functions are

1. *Inclusion*, which regulates motion and/or communication outwards from the interior of a region to the exterior.
2. *Exclusion*, which regulates motion and/or communication inwards from the exterior to the interior.
3. *Separation*, which combines inclusion with exclusion.
4. *Contact*, the extent to which separation is not complete.

Some other functions such as *protection* are derivative from these. There remain functions such as *differentiation*—by which, for example, the land-use within a conservation area is differentiated from that outside even though there may be no restriction to movement or communication across the border in either direction.

A linear feature may be regarded as a boundary to the extent that it embodies one or more of these functions. As already noted, being a boundary is not incompatible with being a way or path; though students of international boundaries repeatedly assert that rivers do not make good boundaries, in part precisely because they make such good thoroughfares (Boggs, 1940; Peattie, 1944; Jones, 1945; East, 1965; Crone, 1967).

In three-dimensional space, boundaries take the form of *surfaces* or *interfaces*. They are intrinsically two-dimensional, and the regions or objects they bound or separate are three-dimensional. Although geographical space—or at any rate 'naive geographic space' (Egenhofer and Mark, 1995)—is generally thought of as two-dimensional, a three-dimensional view is sometimes essential for a true depiction of some geographical state of affairs. Jones (1945) discusses boundary issues arising in relation to mining rights and underground water resources. He cites (p.31) the case of coal mines on the Germany-Netherlands boundary in 1939. The political boundary at the surface was marked by a meandering river course, but the mines underground were allocated to the state in which the coal was brought to the surface, in accordance with a 'working boundary', ratified by treaty, and in places separated from the political boundary by as much as 1km. Such examples contradict the usual assumption that a boundary on the surface of the earth should be regarded as extending vertically upwards and downwards for the purpose of assigning sovereignty to portions of the atmosphere or the earth's crust.

Likewise, although a geological map can only show the boundaries pertaining to the surface geology, the objects of interest to geologists are three-dimensional chunks of matter. For this reason, geological texts often include, as well as conventional surface maps, vertical transects showing the underground distribution of rock types. Even a map of the surface geology carries information about which strata overlie others, and part of the skill of reading such maps is precisely to draw inferences concerning—indeed to visualise—the sub-surface geology. Boundaries in weather maps can similarly only be read correctly on the understanding that they represent two-dimensional boundaries between three-dimensional air masses.

4.2 Valency

As already noted, we may speak both of the boundary *of* one region or the boundary *between* two regions. This may be described as a difference in 'valency'. The issue sometimes arises as to which of the two concepts—the unary 'boundary of' or the binary 'boundary between'—should have conceptual or logical priority. To some extent they are interdefinable. The boundary of England consists of the boundary between England and Scotland, the boundary between England and Wales, and two stretches of coastline representing the boundary between England and the sea. Conversely, the boundary between England and Wales comprises that part of the boundary of England spatially coincident with part of the boundary of Wales, together with that part of the boundary of Wales spatially coincident with the boundary of England. Kulik (1997) invokes this notion of valency to draw the distinction between the German terms *Rand* ('edge') and *Grenze* ('boundary'), modelling the former by means of a one-place function, the latter by a two-place.

The unary conception of boundary plays a prominent role in recent work by Barry Smith and various co-authors, inspired in part by the work of Brentano on continuity in space and time. For Brentano, the essence of continuity is that a continuous expanse can be divided in thought into two pieces whose boundaries are spatially coincident along their line of contact. According to Smith, this picture applies to geopolitical boundaries: 'The boundary of France is not also a boundary of Germany: each points inwards towards its respective territory' (Smith, 1995). This contrasts with the discontinuity between an object and the empty space in which it is located: the common boundary here belongs only to the object. Smith and Varzi (1997) advocate two complementary boundary theories: a bivalent theory to handle physically motivated ('bona fide') boundaries, and a univalent theory to handle cognitively motivated ('fiat') ones.

Against this, we may observe that in the geopolitical case, it usually takes two to make a boundary. The boundaries between nations are generally defined by means of a treaty; or if unformalised, by a truce or armistice. The two parties come to an agreement as to how their mutual boundary should be defined. This is primarily a boundary *between* two regions, and only secondarily part of the boundary *of* either region individually. If a nation A has frontiers with nations B, C, D, ..., then each boundary segment A-B, A-C, A-D, ..., might be defined by a separate treaty. Where three boundaries meet at a triple point, three separate treaties are involved. As Jones says: 'In view of the notorious slowness of boundary negotiations, it is not surprising that overlapping claims in the vicinity of a triple point may exist for many years' (Jones, 1945, pp.160f).

Sometimes, to be sure, a boundary is acknowledged by only one of the two parties involved: an oft-cited example is the boundary between 'East Germany' and 'West Germany' prior to reunification, a boundary officially held to exist only by the East Germans. Such examples do not weaken the earlier argument: it is not a case of there being a single world view in which there is a one-sided boundary of East Germany that does not coincide along its western parts with any one-sided boundary of West Germany; on the contrary, what we had was *two* conflicting world views, according to one of which there was a two-sided boundary dividing Germany into two parts, while according to the other no such boundary existed.

We may constrast, on the one hand, a situation such as we find almost everywhere in the modern world, in which each piece of land is regarded as the sovereign territory of some nation or other, and, on the other hand, the situation which existed in earlier times in which the notion of territorial sovereignty was not yet fully developed, with human groups pushing their frontiers outwards into unclaimed virgin territory beyond. At this stage, each enclave of humanity has a border (albeit seldom precisely defined) representing the limit of its current expansion: not a boundary between two sovereign territories but a boundary of a single territory. One can, of course, think of this as a binary boundary (the boundary between a piece of claimed territory and the wilderness outside) but it seems more natural to think of it as unary. It is when continued expansion brings the frontiers of two such territories into contact that conflict ensues and the necessity arises for establishing a common boundary by mutual agreement. At that point the two unary boundaries are replaced by a single binary boundary. Historically, this has happened at different times in different parts of the world, the process not becoming completed until the twentieth century. Even now, one might argue that coastal boundaries are of the unary kind, and will remain so until such time as the nations decide to partition between them the entire area of the oceans.

In general, where the function of a boundary is to support an exhaustive and exclusive partitioning of an area of land, it is appropriate to regard it as bivalent. Not all boundaries are of this kind, however; in particular not all institutional boundaries are. The national parks of any country are in general isolated from one another, like islands, and while the boundary of a national park separates it from the land outside, this function is asymmetrical. It is not usual to recognise the totality of land not falling within a national park as a single geographical entity; thus it is not natural to think of the boundary of a national park as also being part of the boundary of anything else. It is unequivocally a unary boundary, more like the surface of an orange than the boundary between two counties.

4.3 Determinacy

Important problems arise from the fact that boundaries are often ill-defined in various ways. Certain types of boundaries are susceptible, in principle, to precise definition. With institutional boundaries, the underlying intention is to define an exact Euclidean line, having length but no breadth. If this intention does not always succeed it may be because of the difficulties in securing an exact delimitation in practice, or because of ambiguities in the wording of document by which the boundary is defined. Again, the isolines of a field whose variation is truly continuous are, in principle, absolutely sharp, any apparent indeterminateness arising from limitations in the accuracy with which we can measure the relevant values. Heterolines, too, will be sharp to the extent that the attributes in terms of which they are defined are precisely determined at each point.

Material boundaries, considered as boundaries, are not sharp in this sense; and since a material boundary is also a region, a further issue arises as to whether its own boundaries are sharp. This leads to higher-order indeterminacy. To revert to an earlier example, the *Kampfzone* is a non-sharp boundary between forested and treeless regions; since it is a transition zone with a distinctive character of its own, its non-sharpness is not due to problems of measurement or definition, rather it is in the nature of things that the regions it separates do not have a sharp boundary. The *Kampfzone* itself has boundaries: the timberline and treeline. These are also not sharp, but this is not because they are themselves transition zones, but because of it is impossible to specify them more narrowly than the size of a typical tree (or the gap between trees).

A major source of indeterminateness arises from attempts to define the boundary of an object which, properly speaking, has no boundary. A well-defined location for a mountain is provided by its summit. But can we encircle the summit with a line enclosing all and only those places that form part of the mountain? One can invent criteria, which may result in quite sharp delineations, but none of them corresponds to our ordinary understanding of what a mountain is—which perhaps includes boundarylessness as a significant attribute. Unfortunately, many representational tools cannot assign a location to anything without assigning a sharply-defined location.

One way to address problems of this kind is by means of inner and outer boundaries, effectively creating a notional transition zone in between. This may or may not be a true transition zone defined in terms of the transition of pre-existing characters. The limits of the transition zone may be more or less arbitrary, the only essential requirement being that everything inside the inner boundary unequivocally belongs to the region whose delineation is in question, and everything outside the outer boundary is unequivocally outside it. Formal developments of this approach are presented by Clementini and Di Felice (1996) and by Cohn and Gotts (1996). More generally, see other chapters in Burrough and Frank (1996) for discussions of indeterminate boundaries in geography.

5   REPRESENTATION OF BOUNDARIES

Suppose we wish to construct a model—mathematical, conceptual, or computational—of part of geographical space. What kinds of boundaries should we represent, and how should we represent them? In the absence of a more specific context, this question has no definite answer. We need to know more about the purpose of the model, and the resources it can draw upon. Here I organise the discussion in terms of two distinct approaches to modelling geographical information. At the conceptual level, appropriate for the initial design of an information system, these are the *field-based* and *object-based* approaches; at the level of implementation, they show up as *raster-based* and *vector-based* GIS.

In the field-based approach, data are presented in the form of *fields*, which are functions assigning values to spatial locations. An example is the elevation field, which assigns to each point on the earth's surface the elevation of the solid surface above or below sea level. When data are presented in this way, there is no natural way of representing boundaries explicitly; but this does not mean that the data cannot be used to derive information about the location of boundaries. The kinds of boundaries we can retrieve depend on the nature of the field data and the underlying spatial framework. This may be continuous or discrete. In the former case it is naturally modelled mathematically by means of tuples of real numbers, in the latter by tuples of integers. Moreover, the field values may also be either continuous or discrete, modelled by reals, integers, or subranges thereof. I shall consider various possible combinations in turn:

Continuous space, continuous field values   The *variation* in the field values over space may itself be either continuous or discontinuous. In the former case the natural boundaries are isolines. These are explicitly present in the model in the sense that, for example, the $x = k$ isoline consists of a set of points to which the x-field assigns the value k, and these points, and the assignment of this value to them, are explicitly present; but the model does not in itself draw attention to the points lying along any particular isoline as opposed to the infinitely many alternative possibilities, and in that sense the isolines in such a model are only implicit. This holds even at an idealised conceptual level; any actual implementation must represent the continuous field by means of some finitely-specifiable approximation. Here the isolines may be even less explicit, since interpolation may be required to estimate their courses. It should be stressed that while any isoline *can* be regarded as a boundary, isolines do not in themselves constitute boundaries. A continuous field-based model does not in itself provide a means for us to designate this or that isoline as a boundary.

Similarly, there may also be transition zones present in the field data, but their presence is not explicitly shown by any mechanism within the model itself: the model merely provides the raw data on the basis of which the boundary may be defined. If the variation in the field values is discontinuous, we will find heterolines marking the discontinuities. Separation zones of both kinds may exist with both continuous and discontinuous fields.

Continuous space, discrete field values   There is a familiar problem arising in this case. Suppose v and w are two different values of the field, and suppose the area with value v meets the area with value w along a line L. Which value should be assigned to the points along L? This depends on the nature of the discrete field. The easiest case is when the field has arisen by a discretisation of some continuous field f. A typical case is where the values v and w are defined as, say, $f < k$ and $f \geq k$ respectively. In this case, if f is truly continuous, the points along L must take the value w, since f must equal k at those points (see Galton, 1997, for a discussion of the temporal analogue of this case). If the discrete values v and w are more qualitative in nature, this type of solution is unavailable, and now it seems arbitrary which of the values is taken. An example of this, discussed by Casati and Varzi (1999), is Peirce's puzzle: what colour is the line of demarcation between a black spot and its white background? The puzzle arises as an artefact of the modelling process. In reality the attribute of 'colour' applies to areas, not to individual points. In a continuous field-based model, however, all spatial attributes are ascribed to points; they are ascribed to regions only indirectly via the points which fall within them. In a common-sense view of reality one wants to be able to say that a white area meets a black area without having to ascribe a colour to the points along the boundary; this cannot easily be accommodated in a discrete field defined on a continuous space.

Discrete space, discrete field values In this case the natural boundaries are heterolines. The discrete space consists not of points but of minimal units of area ('cells'), linked by a relation of adjacency. If a block of black cells meets a block of white cells then the hetero-line representing the boundary between black and white does not itself consist of cells but rather follows a line of interstices between pairs of adjacent cells. It is an epiphenomenon of the representation, truly reflecting the epiphenomenal character of the boundary in the represented reality. Likewise, a discrete field model may contain separation zones, for example a band of black cells separating areas of blue and green. And as with transition zones in a continuous field, a separation zone in a discrete model is explicit in the sense that it can be identified with an actual collection of cells distinct in value from those on either side, but merely implicit in the sense that the model does not in itself draw attention to this particular heterogeneity by collecting cells together into a unity.

An important characteristic of field-based models to emerge from this discussion is that in such models, boundaries are typically represented *analogically*: that is, a boundary in reality shows up as a boundary in the model—an isoline in reality is represented by an isoline in the model, and likewise with heterolines, transition zones and separation zones. In all these cases, although the brute facts concerning the physical nature of the boundary are explicitly represented, the boundary itself, as a geographical entity with its own particular properties, seems curiously absent. This reflects the fact that field-based models provide rather low-level representations of reality. They are particularly appropriate for handling data collected directly from nature using automatic sensors, for example. But boundaries, as part of our conceptual scheme, seem to belong at a somewhat higher level. This higher-level conceptual character can be more explicitly captured by means of an object-based representation.

In an object-based model, the primary data consist of conceptual elements called *objects*, to which are assigned various *attributes*. Spatial location is just one amongst many possible attributes that can be assigned to objects in such a model. There is no reason why boundaries should not be explicitly represented as objects in their own right. And indeed, it is standard practice, in vector GIS, to specify the location of an area object by specifying the location of its boundary, using the coordinates of a suitably chosen sequence of points located along the boundary.

Any boundary can, in principle, be represented in an object-based model. A boundary is handled like any other linear feature; its location specified as a sequence of points. But being an object, it can have any number of other attributes assigned to it as well: whether it is a physical or institutional boundary, and if physical, whether material or epiphenomenal, and so on. Institutional boundaries can be provided with exact characterisations of their type: whether it is an international boundary, a state boundary, a county boundary, a parish boundary, the boundary of an electoral constituency, and so on. With international boundaries one could distinguish those that are ratified by treaties and those which are merely *de facto*; one could provide information about when the boundary came into being, and so on. None of this is possible in a purely field-based model.

In contrast to field-based models, the representation of boundaries in an object-based model is *symbolic* rather than analogical. That a boundary is an isoline, heteroline, transition zone, etc., is not shown analogically by the fact that the representation shares that character—it does not—but rather by means of particular symbols conventionally understood as representing the various kinds of boundary. On a map, these show up as the many different styles of broken or coloured lines that may be used to represent different forms of boundary—the non-analogical character of such representations is evident in the fact that without consulting the legend it is often not possible to tell which of the many linear features on a map represent boundaries and which represent paths, railways, etc.

Another consequence of the symbolic character of object-based boundary representations is that the issue of indeterminateness becomes highly problematic. In a field-based model, a transition zone can be represented analogically *as* a transition zone, with the intermediate field values directly representing the gradation in the underlying reality. In an object-based model, on the other hand, everything is biased towards a crisp all-or-nothing style of representation, and in order to represent indeterminate boundaries one has to have recourse to artefacts such as the 'egg-yolk' theory of Cohn and Gotts (1996), which models the zone of indeterminacy as a kind of region in its own right, itself with crisp boundaries.

An object-based model is a higher-level representation that is hard to extract mechanically from raw data, involving as it does a conceptualisation imposed on those data by means of human thought processes. Since institutional boundaries essentially involve such higher-level aspects, these boundaries are particularly apt to be represented as objects. While of course it is possible to represent, say, political or administrative regions by means of field values—in which case their boundaries will show up as heterolines in the representation—this does not seem very natural. In Galton (2001) I propose a hybrid model—the 'object-field'—as a way of handling the partition of a land area into institutionally-defined blocks.

In conclusion we may say that the two styles of geographic modelling both provide the ability to represent boundaries, but with very different advantages and disadvantages. To the extent that boundaries are themselves physical features, they can be represented analogically within a field-based representation, which may faithfully reflect various physical characters of different kinds of boundary (material vs epiphenomenal, crisp vs fuzzy, etc.); but these representations do not draw attention to the boundaries themselves. The boundaries do not resolve themselves into discrete objects that we can readily identify and talk about. In object-based models, on the other hand, boundaries can be elevated to the status of objects, to which we can ascribe, by symbolic means, whatever properties we wish; in particular, institutional boundaries, which are not directly grounded in the sorts of properties most readily represented in a field-based model, are much more comfortably accommodated in the object-based framework.

6 CONCLUSIONS

In this chapter I have described a possible classification of boundary types, and in the light of this classification have considered how boundaries of different kinds can be related to one another, what attributes of boundaries are of particular importance in giving a general characterization of them, and how boundaries and their attributes may be represented in an information system. This work may be regarded as a prolegomenon to future studies in which the representation of boundaries in information systems is considered in more detail, with a view to improving the quality of boundary information that can be provided by such systems. It is to be hoped that much of what is said here may also be of interest to wider communities of geographers, philosophers, and others.

REFERENCES

Boggs, S. W. (1940). *International Boundaries: A Study of Boundary Functions and Problems*. New York: Columbia University Press.

Burrough, P. A. and Frank, A. U. (Eds) (1996). *Geographic Objects with Indeterminate Boundaries*. London: Taylor & Francis.

Casati, R. and Varzi, A. C. (1999). *Parts and Places: The Structures of Spatial Representation*. Cambridge, MA: MIT Press.

Clementini, E. and Di Felice, P. (1996). An algebraic model for spatial objects with indeterminate boundaries. In Burrough, P. A. and Frank, A. U. (Eds), *Geographic Objects with Indeterminate Boundaries*, pp. 155–169. London: Taylor & Francis.

Cohn, A. G. and Gotts, N. M. (1996). The 'egg-yolk' representation of regions with indeterminate boundaries. In Burrough, P. A. and Frank, A. U. (Eds), *Geographic Objects with Indeterminate Boundaries*, pp. 171–187. London: Taylor & Francis.

Collinson, A. S. (1988). *Introduction to World Vegetation*. London: Unwin Hyman, 2nd edition.

Couclelis, H. and Gottsegen, J. (1997). What maps mean to people: Denotation, connotation, and geographic visualization in land-use debates. In Hirtle, S. C. and Frank, A. U. (Eds), *Spatial Information Theory: A Theoretical Basis for GIS*, volume 1329 of *Lecture Notes in Computer Science*, pp. 151–162. Berlin: Springer-Verlag.

Crone, G. R. (1967). *Background to Political Geography*. London: Museum Press.

Dorling, D. and Fairbairn, D. (1997). *Mapping: Ways of Representing the World*. Addison Wesley Longman.

East, W. G. (1965). *The Geography Behind History*. London: Nelson.

Egenhofer, M. J. and Mark, D. M. (1995). Naïve geography. In Frank, A. U. and Kuhn, W. (Eds), *Spatial Information Theory: A Theoretical Basis for GIS*, volume 988 of *Lecture Notes in Computer Science*, pp. 1–15. Berlin: Springer-Verlag.

Fielding, G. J. (1974). *Geography as Social Science*. New York: Harper & Row.

Galton, A. P. (1997). Space, time and movement. In Stock, O. (Ed), *Spatial and Temporal Reasoning*, pp. 321–352. Dordrecht: Kluwer Academic.

Galton, A. P. (2001). A formal theory of objects and fields. In Montello, D. R. (Ed), *Spatial Information Theory: Foundations of Geographical Information Science*, volume 2205 of *Lecture Notes in Computer Science*, pp. 458–473. Berlin: Springer-Verlag.

Harley, J. B. (1975). *Ordnance Survey Maps: A Descriptive Manual.* Southampption: Ordnance Survey.

Hurvich, L. M. (1982). *Color Vision.* Sinauer Associates Inc.

Jones, S. B. (1945). *Boundary-Making: A Handbook for Statesmen, Treaty Editors and Boundary Commissioners.* Washington DC: Carnegie Endowment for International Peace.

Kulik, L. (1997). Zur Grenzziehung zwischen Rand und Grenze. In Krause, W., Kotkamp, U., and Goertz, R. (Eds), *KogWis97. Proceedings der 3. Fachtagung der Gesellschaft für Kognitionswissensschaft*, pp. 104–106, Jena. Friedrich-Schiller-Universität.

Lewis, J. R. (1964). *The Ecology of Rocky Shores.* London: English Universities Press.

Mark, D. M. and Csillag, F. (1989). The nature of boundaries on 'area-class' maps. *Cartographica*, 26:65–78.

Pears, N. (1977). *Basic Biogeography.* London: Longman.

Peattie, R. (1944). *Look to the Frontiers: A Geography for the Peace Table.* New York: Harper & Brothers.

Pielou, E. C. (1979). *Biogeography.* New York: Wiley.

Plewe, B. (1997). A representation-oriented taxonomy of gradation. In Hirtle, S. C. and Frank, A. U. (Eds), *Spatial Information Theory: A Theoretical Basis for GIS*, volume 1329 of *Lecture Notes in Computer Science*, pp. 121–135. Berlin: Springer-Verlag.

Prescott, J. R. V. (1965). *The Geography of Frontiers and Boundaries.* London: Hutchinson.

Putman, R. J. (1984). The geography of animal communities. In Taylor, J. A. (Ed), *Themes in Biogeography*, pp. 163–190. London: Croom Helm.

Searle, J. R. (1995). *The Construction of Social Reality.* Harmondsworth: Penguin.

Shalowitz, A. L. (1962). *Shore and Sea Boundaries.* Washington DC: US Department of Commerce.

Smith, B. (1995). On drawing lines on a map. In Frank, A. U. and Kuhn, W. (Eds), *Spatial Information Theory: A Theoretical Basis for GIS*, volume 988 of *Lecture Notes in Computer Science*, pp. 475–484. Berlin: Springer-Verlag.

Smith, B. and Varzi, A. C. (1997). Fiat and bona fide boundaries: Towards an ontology of spatially extended objects. In Hirtle, S. C. and Frank, A. U. (Eds), *Spatial Information Theory: A Theoretical Basis for GIS*, volume 1329 of *Lecture Notes in Computer Science*, pp. 103–119. Berlin: Springer-Verlag.

Tivy, J. (1971). *Biogeography: A Study of Plants in the Ecosphere.* Edinburgh: Oliver & Boyd.

Walter, H. (1985). *Vegetation of the Earth and Ecological Systems of the Geo-biosphere.* Berlin: Springer-Verlag, 3rd edition.

Worboys, M. F. (1995). *GIS: A Computing Perspective.* London: Taylor & Francis.

Yonge, C. M. (1949). *The Sea Shore.* London: Collins.

CHAPTER 9

Regions in Geography: Process and Content

Daniel R. Montello
Department of Geography, University of California
Santa Barbara, CA, 93106, USA

1 INTRODUCTION

The concept of regions has nearly always been of central importance to geography and other disciplines that study earth-referenced phenomena (Linton, 1951; Kostbade, 1968; Richardson, 1992; Martin and James, 1993). From the Classical period to the Modern and Postmodern, the identification, description, and explanation of regions has played a critical role in attempts to understand and control the earth and its phenomena. "Understanding the idea of region and the process of regionalization is fundamental to being geographically informed" (Geography Education Standards Project, 1994, p.70). Although geographers' relative emphases on idiographic description of unique places versus nomothetic explanation of abstract truths has varied greatly over time (and still varies greatly over university space), regionalization understood broadly has remained important. A travel log describing the cultures and climates of distant lands organizes the earth into regions; so too does a computational model of migration flows based on census data (see Hargrove and Hoffman, 1999, for evidence of the continued importance of regions to quantitative and scientific geography). Regionalization operates somewhat differently in regional and systematic (special and general) approaches to the study of geography, but regionalization is important either way. Even today, there is warrant to the traditional claim (Kimble, 1951) that geography is the "study of regions."

In this essay, I revisit the regional concept in the light of some recent philosophical and technical advances that have been made in geography. I contrast the concept of geographic regions with that of regions generally and of spatial objects even more generally. What makes geographic regions a cogent idea, related but not identical to other entities that make up a complete ontology of space and place? To refine this ontology, I propose a taxonomy of geographic regions based on their contents and the processes by which they are formed. The taxonomy consists of four types: administrative, thematic, functional, and cognitive regions. These four types represent an elaboration and clarification of existing taxonomies, such as those discussed in the literature of geography, including many textbooks. Among other refinements, I differentiate formal regions into administrative and thematic regions, and explicitly consider the individual vs. social nature of cognitive regions. I then consider the important issue of boundaries, including boundary vagueness and how it applies to the four region types. The special status of administrative regions is noted, and the ability of Smith's fiat/bona-fide distinction to account for them is examined. I consider many of the ontological issues surrounding regions as part of an apocryphal lunchtime conversation over one of the classic and still enduring puzzles of philosophy—the mind-body question. The essay concludes by arguing for the continued meaning and relevance of the region concept in the current and future world of digital geography.

1.1 What are geographic regions?

Regionalization, the creation or identification of regions, is a subset of *categorization* (Kostbade, 1968; Goodchild, 1992). Categorization is the identification of discrete sets of entities, physical or conceptual. Categories delimit entities which share one or more properties from entities which do not share the properties. These"properties" vary widely in their characteristics—their abstractness, their functional importance, their conceptual importance, and so on (Smith and Medin, 1981). The essential aspect of a category is that it contains entities grouped together as similar in some way and distinguished from entities in some other category. Thus, at minimum, the shared property may simply be that entities in the category have been capriciously assigned to have the property "members of that category"; other entities have been assigned the property "not in that category." Such a minimal property is not often of interest but does serve to establish a baseline definition for categorization. These considerations also hold for spatial categories—regions. The necessary and fundamental spatiality of regions distinguishes them from most other category systems. Other category systems are often spoken about or depicted "spatially," but this is metaphorical. For example, the words *pretty, beautiful*, and *attractive* may be depicted as located in a common region of a space of word meanings, but that is not a literal region nor a literal space but a metaphorical "semantic space" (hence the quote marks).

I focus in this essay on *geographic regions*, pieces of (near) earth surface. Geographic regions, as a subset of regions more generally, have certain shared properties. Spatially, they are prototypically 2-dimensional, at or near the earth surface, though exceptions exist (e.g., volumetric regions underground or in the atmosphere are sometimes identified). Geographic regions are usually (but not always) defined not only spatially but according to what is there, the "content" or "theme" of the region. The contents are human and natural entities or processes, the thematic components of geography's subject matter. If thematic dimensions are added to the spatial dimensions, geographic regions are always 3 to N dimensional. Also characteristic of geography, as the "study of the earth as the home of humanity" (Martin and James, 1993), is its concern with phenomena at some spatial and temporal scales more than at others (Montello, 2002). Although advances have occurred and will continue to occur when geographers stretch the scale of their subject matter, few would argue that molecular or interplanetary scales are properly of concern for geography. But aside from this modest (from the perspective of human experience) scale restriction, I do not favor the traditional constraining concept of a specific "regional scale" of analysis (e.g. Meyer et al., 1992) when considering the fundamental ontology of regions. Such a regional scale would restrict our focus to areas that are something like larger than local communities and smaller than continents. Nor does it seem principled in this context to distinguish sharply between *places* and *regions* (e.g. Richardson, 1992); the first may be considered a subset of the latter.

Geographic regions are thus examples of spatial regions in general. Similarly, geographic regions are also instances of geographic *features* or *objects*. At small cartographic scales (i.e., large areas of earth), some small regions need to be understood as point-like or line-like features rather than extended pieces of earth surface. The application of the concepts of regions and objects as abstract ontological entities (Smith and Varzi, 1997; Casati and Varzi, 1999) to the concept of geographic regions is an important and ongoing concern for geographic information scientists (e.g. Frank et al., 2001; Mark et al., 2001). Even given my focus on geographic regions, as opposed to regions and geographic features more generally, the conception of geographic regions I consider here is broader than that discussed by many writers in the geographic literature. Some, for instance, consider regions solely as the expression of possibly discontinuous natural and human reality (Kimble, 1951); others focus exclusively on regional creation as an expression of socially-constructed reality (Aitken, 2001). The taxonomy I introduce below incorporates both of these conceptions and more.

Geographic regions (henceforth *regions*) need not be contiguous, wholly interconnected pieces of earth area, but usually they are. In fact, not only are they usually unfragmented but fairly compact, not very elongated or prorupt in shape. There are several reasons why regions are usually spatially contiguous and compact areas. One is that structures and processes are not randomly distributed over the earth—they cluster together, reflecting *spatial autocorrelation*. Spatial autocorrelation on a fairly isotropic 2-dimensional surface creates compact regions. A second reason for compactness is that people tend to group things together that are closer together in space (and time)—there is a perceptual and cognitive penchant to see the world this way, as proposed in the organizing principles of Gestalt psychologists such as Wertheimer in the early 20th C. (Kaufman, 1974). Yet a third reason for contiguous and compact regions is their utility. They help people organize their understandings of the world in an efficient manner; they also help various activities in space occur more efficiently. A good example of the consequences of ignoring the latter rationale is provided by administrative regions that are quite noncontiguous or noncompact. They force complex and repeated boundary crossings, or indirect and expensive networks of interaction, as part of activities like commerce or migration. Such irregular regions are typically created to serve a political purpose that intentionally violates "geographic rationality" in order to serve some other rationality. A fascinating case in point is racial gerrymandering of voting districts in the U.S. (Forest, 2001), where courts have sometimes recognized geographic rationality by ruling against "bizarre" districts that are extremely noncompact. Of course, regional fragmentation is sometimes geographically rational insofar as it gives locational presence to a political entity at the farthest reaches of its desired territorial influence; Hawaii and Alaska both provide such a benefit to the U.S., and thus rationally constitute part of the country in spite of the high costs of living (and the cartographic nuisance) engendered by their noncontiguity.

1.2 Regions in thought

Above, I referred to the centrality of region concepts in geography, historically and currently. We can go further—there is every reason to believe that the first geographic musings of our Paleolithic brothers and sisters made plentiful use of regional thinking (particularly of the cognitive variety discussed below). That is because regionalization, and categorization in general, universally characterizes human thought. Humans organize knowledge categorically. Neither logic nor evidence suggests there is any place or time where this does not hold, at least as a characteristic of "common sense" cognition (see, for example, any historical or anthropological text that documents the universal characteristic of humans to label geographic features in the world, including regions). Furthermore, categorization is universal though it produces distortions in thought. We group partially similar things together, ignoring many differences. We distinguish between things that share many similarities. In doing this, we minimize intra-category variation and exaggerate extra-category variation (see Tajfel and Wilkes, 1963, and the copious research it has generated in the areas of decision-making and stereotyping). Many of us are familiar with the inflated distinctions border residents draw between themselves and "those people over there." An apocryphal tale recounted by Muehrcke and Muehrcke (1992) nicely exemplifies categorical reasoning about regions. A man living near the Canadian-Alaskan border wasn't sure on which side of the border he lived. So he hired a surveyor, who determined that the man lived in Canada. "Thank God!", the man cried, "now I won't have to live through another of those terrible Alaskan winters!"

Anecdotal and apocryphal evidence aside, many empirical studies have demonstrated the regional organization of geographical knowledge, including its effects on judgments, whether based on maps or direct experience (Stevens and Coupe, 1978; Maki, 1981; Hirtle and Jonides, 1985; McNamara et al., 1989; Friedman and Brown, 2000). It appears clear that humans think in discrete pieces of truth or reality, even if that way of thinking is in fact false or distorted. But like categorization in general (Smith and Medin, 1981), regionalization has its definite analytic and communicative utility. It simplifies complexity and avoids unnecessary precision, both in thought and in speech (Talmy, 1983; Freksa, 1991; Landau and Jackendoff, 1993). It contributes greatly to the efficiency of our interaction with the world, making it unnecessary to learn properties of objects and events anew each time we encounter another instance. Discretization of the world may be especially valuable because it allows us to integrate or combine the separate views of the world we experience from our local perspectives into internal representations that go beyond local truth (Tversky, 2002).

2 PROCESS- AND CONTENT-BASED TAXONOMY OF REGIONS

A great deal has been written about regions in geography, including various taxonomies of regions. In this section I review traditional taxonomies of region types, and propose a new taxonomy based on the content of regions and the processes by which they are formed. The taxonomy I propose refines some traditional distinctions commonly made among types of regions (see below) and provides more meaningful labels for the types. The taxonomy consists of four types: administrative, thematic, functional, and cognitive regions. *Administrative regions* are formed by legal or political action, by decree or negotiation. These include regions based on property ownership (cadastral regions) and on political and administrative control, such as census tracts, provinces, and countries. *Thematic regions* are formed by the measurement and mapping of one or more observable content variables or themes. They show where some entities or properties exist; the entities may be natural (rainfall, pine trees) or human (languages, crops) in origin. *Functional regions* are formed by patterns of interaction among separate locations on the earth. Spatial interaction is fundamentally the movement of matter or energy from place to place: people, commodities, water, seeds, earthquake tremors. Pattern in energy flows can encode information, the basis for communication, which is thus a form of spatial interaction. Finally, *cognitive regions* are produced by people's informal perceptions and conceptions (downtown, the Midwest).

It is useful to recognize explicitly that both thematic and functional regions may be based on single variables or multiple variables—they may be *univariate* or *multivariate* regions (Burrough, 1996 refers to *polythetic* regions). A univariate thematic region is based on a single theme, such as average temperature; a multivariate thematic region is based on combinations of themes, such as culture regions or ecological biomes. Similarly, a univariate functional region is based on a single process of interaction, such as seed dispersal; a multivariate functional region is based on combinations of interactions, such as media reception that includes radio, newspaper, and television. The distinction between univariate and multivariate regions is not often made explicitly (e.g., in textbooks), but it is very important. For one, multivariate regions are potentially vague in ways that univariate regions are not, as discussed below. However, one should distinguish the idea of multivariate regions from the idea that any type of region may be identified differently at different spatial scales, whatever the basis for regionalization. These different region sets often overlap in space, producing the popular geographic notion of a hierarchy of regions based on scale (e.g. Golledge, 1992). Countries contain provinces, climates contain micro-climates, downtown contains subregions varying in hipness.

As stated above, several region taxonomies have been proposed by geographers in the past. A very influential taxonomy was proposed by Hartshorne (1959), who distinguished *formal, functional,* and *general* regions (I describe these below). Many textbooks in both human and physical (natural) geography present taxonomies of regions quite similar to Hartshorne's (Table 1 lists several representative texts). All texts liberally use the concept of region. Nearly all explicitly define region. Like my definition above, the texts generally refer to the internal similarity and external dissimilarity of regions. It may be significant, however, that texts of human geography are much more likely to elaborate the concept of regions by describing a taxonomy of two or three types than are texts of physical geography. Here are some likely reasons for this:

a. Administrative regions are clearly human creations. Although physical features often help mark these boundaries, they are generally not the boundaries, and in no sense does the administrative region exist in the least without humans (if a national border is crossed and no agents are there, does it arouse anti-immigrant sentiments?)—thus no administrative regions in physical geography. Of course, administrative regions have great consequences for human management of the natural world, but that has not traditionally been a concern of physical geographers.

b. Physical geographers believe that as scientists, only "objectively" measurable constructs may be countenanced—thus no cognitive regions. As the Accountant and the Swineherd discover below, this is an overly dualistic attitude that falsely impugns beliefs as unreal, exaggerates the objective-subjective distinction, and fails to recognize that natural scientists (in many disciplines anyway) regularly think and communicate in terms of cognitive natural regions, though they may hold up the eventual objectification of the region as an ideal. Bog lands, for example, are frequently referred to without formal measurement or precise definition.

c. Spatial interaction is defined, at least in human geography texts, as referring to *intentional* contact among people, including transportation, communication, and commodity exchange—thus no functional regions in physical geography. But many instances of human interaction are not intentional. Furthermore, contact among separated locations via the movement of matter and energy occurs constantly in the natural world, and plays a central role in the ideas of natural scientists; Kostbade (1968) identified river systems as functional regions.

Human

Bergman, E. F. (1995). *Human Geography: Cultures, Connections, and Landscapes.* Englewood Cliffs, NJ: Prentice Hall.

de Blij, H. J. (1999). *Human Geography: Culture, Society, and Space.* New York: John Wiley & Sons, 6th edition.

Fellmann, J., Getis, A., and Getis, J. (1995). *Human Geography: Landscapes of Human Activities.* Dubuque, IA: Wm. C. Brown Publishers, 4th edition.

Knox, P. L. and Marston, S. A. (1998). *Places and Regions in Global Context: Human Geography.* Upper Saddle River, NJ: Prentice Hall.

Kuby, M., Harner, J., and Gober, P. (1998). *Human Geography in Action.* New York: John Wiley & Sons.

Rubenstein, J. M. (1996). *The Cultural Landscape: An Introduction to Human Geography.* Upper Saddle River, NJ: Prentice Hall, 5th edition.

Physical

Christopherson, R. W. (1997). *Geosystems: An Introduction to Physical Geography.* Upper Saddle River, NJ: Prentice Hall, 3rd edition.

de Blij, H. J. and Muller, P. O. (1996). *Physical Geography of the Global Environment.* New York: John Wiley & Sons, 2nd edition.

Gabler, R. E., Sager, R. J., and Wise, D. L. (1993). *Essentials of Physical Geography.* Fort Worth: Saunders College Pub, 4th edition.

Marsh, W. M. and Dozier, J. (1981). *Landscape, an Introduction to Physical Geography.* Reading, MA: Addison-Wesley.

Wallen, R. N. (1992). *Introduction to Physical Geography.* Dubuque, IA: Wm. C. Brown Publishers.

Table 1: Introductory geography textbooks examined for this essay

Likely following Hartshorne, geography textbooks that offer region taxonomies all differentiate formal and functional regions. A *formal region* is defined as "distinguished by a uniformity of one or more characteristics" or "an area in which the selected trend or feature is present throughout." Examples given include soil regions, dialect regions, topography regions, and often, political regions like Ecuador. This concept of formal regions, and its definitions, is problematic. Formal regions include what I call thematic regions, but many writers also include as formal regions what I call administrative regions. In fact, thematic and administrative regions share very little conceptually. Administrative regions are an important and distinct type, and certainly deserve separate recognition. Failing to recognize administrative regions as a type is acceptable in a dehumanized earth science but not in geography. But setting aside the lack of recognition of administrative regions, what is meant by saying that climate or language regions are "uniform" or "homogeneous?" Even univariate thematic regions are rarely uniform throughout with respect to the relevant theme. Multivariate thematic regions are anything but uniform or homogeneous. The coastal sage-scrub ecosystem contains characteristic plants, rainfall, soil, and distance from the ocean to varying degrees across its extent. Furthermore, there is not complete agreement among experts as to what and how themes should be combined in order to define this ecosystem. The standard definition of formal regions in geography seems quite misleading.

Again following Hartshorne, *functional regions* are "defined by the particular set of activities or interactions that occur within it" or "an area in which an activity has a focal point." This is essentially the same functional region I include in my taxonomy. Definitions of functional regions frequently offer *nodal region* as a synonym. The idea here is that interaction emanates from a node, a point-like feature from which movement comes. This is often true but not always. The region that receives sirocco winds in Mediterranean Europe is a functional region, but the winds emanate from a fairly large area in North Africa, not a point. The region of the U.S. that receives fruits and vegetables from California has a similarly non-punctate source.

Hartshorne listed *general region* as his third type, which means something like a region that "stands out in one's mind" or is "distinctive." This type label has not found widespread usage in geography, probably because its definition as a separate type is not coherent. Instead, texts in human geography often refer to perceptual regions. A *perceptual region* is defined as "a region that only exists as a conceptualization or an idea and not as a physically demarcated entity," "a region perceived to exist by its inhabitants or the general populace," or "an area defined by subjective perceptions that reflect the feelings and images about key place characteristics." First, beliefs, concepts, attitudes, and memories are more properly termed *cognitive* rather than *perceptual*, perception being a subset of cognition that technically refers only to the acquisition of knowledge of local surrounds via the senses. The geographers' broad use of the term *perception*, as in the subdiscipline of "environmental perception," has been questioned before; to be fair, using it this way has precedence in the common-sense metaphorical extension of the word, as in "I see what you mean." Another difficulty is that standard definitions of perceptual regions typically state or imply that such regions are not "real" like other regions are. The Accountant and the Swineherd consider that point below. Finally, an important aspect of many definitions is the degree to which they imply that perceptual regions are socially shared. Examples of perceptual regions always focus on widely held conceptions such as "the South" or "the Bible Belt." In fact, it is useful to recognize that cognitive regions may be idiosyncratic to one or two individuals (such as "our meeting place in parking lot 12") or shared among members of some social or cultural group. When cognitive regions are shared among members of a cultural group, geographers speak of *vernacular regions* as a subset of cognitive regions (for some reason, they never speak of "elitist" cognitive regions). I think the blurring of individual and cultural cognition is unfortunate, though one clearly blends into the other. However, the long-debated question of whether aggregated individual cognition makes culture, or instantiated cultural cognition makes individual minds, is beyond the scope of my discussion here.

Perhaps four region types are unnecessary—maybe fewer will do the job. For instance, one could argue that all regions in geography are actually variants of just one of the four types. Thus, all regions are thematic, insofar as they are regions of earth surface defined by the presence of one or more themes: the region over which the Chinese government claims sovereign rights, the region where a particular newspaper is read, the region that people say is their homeland. On the contrary, all regions are functional. Libya represents the migratory and military interaction of Libyan peoples over space and time; the high desert region results from the movement of continental plates, winds and clouds, and seed dispersals; downtown is the area where tourists go shopping. Finally, it is well known that all regions are cognitive, insofar as every recognition of a region can be considered fundamentally to be a cognitive act, an act of intent or belief. Cognitive systems create realities and include external knowledge technologies such as cluster-analysis software. Thus, one can apparently define all four types of regions in terms of one of the others. The apparent exception is administrative regions; only then does this paragraph's exercise in taxonomic parsimony appear to fail seriously—it does not appear that one could redefine all regions as being administrative. In spite of the success of our efforts at parsimony, though, I believe there is value in making taxonomic distinctions among region types like those I propose in this essay. The ability to use thematic, functional, and cognitive regions to subsume the other types argues against the favored status of one of them. I believe a reduction into one or two types would obscure important process and content distinctions among thematic, functional, and cognitive regions. The difficulty of trying to define all regions as administrative regions points to their special status as a type, and the issue of boundaries we consider in the next section provides further rationale for recognizing the uniqueness of administrative regions.

One final point in this section deserves comment. In applying the taxonomy, it is critical to recognize that people use the same region label at different times to refer to different regions; they also use it to refer to different *types* of regions. This is common. For example, *California* is most often thought of as an administrative region but could also be a thematic region, a functional region, or a cognitive region (a "state of mind"). As an administrative region, Bakersfield is just as much in California as is Los Angeles. But for most people, Los Angeles is more clearly in the cognitive region of California than is Bakersfield. As another example, *France* suggests everything from nasal vowels to fine cuisine, but as a thematic, functional, or cognitive region, not as an administrative region. I do not assume there is anything unique about this polysemy to the region concept; any concept of great richness and wide applicability probably displays this flexibility.

3 INSIDE, OUTSIDE, AND THE GREAT DIVIDE

Regions may be considered from two perspectives: what is inside or outside the region, and what is the divider between inside and outside. Of course the two perspectives are complimentary, necessarily implying each other. But the perspective of the divider or *boundary* is the more problematic of the two and has attracted a great deal of attention in the last couple decades from several of the disciplines that make up geographic information science. For one, boundaries are prototypically thought of as lines but are often not lines. That is, they are 2-dimensional rather than 1-dimensional. The 2-dimensionality of region boundaries is usually due to their vagueness (Mark and Csillag, 1989). *Vagueness* here simply means that a divide between inside and outside a region is not as precise as it could be in theory, so it is a band of nonzero width instead of a (near) geometric curve of little or no width. Common synonyms include *imprecision, indeterminacy, ill-boundedness, gradation, error, uncertainty,* and *fuzziness*. These are in fact only near synonyms. For example, vagueness is not always due to a mistake or error, vague boundaries are often certainly vague, and fuzzy logic provides only one specific model of vagueness (Fisher, 1996). There are cases where the 2-dimensionality of a boundary is not due to vagueness (or any of the related concepts), but these are rare; an example might be the Demilitarized Zone in Korea, a precisely delimited 2-dimensional boundary of uniform status, at least as a political entity. Interestingly, although administrative boundaries are the most precise boundaries of any region type, they may be 2-dimensional because of their extension into the vertical, that is, above and below ground.

One or more of the following reasons can explain why a particular boundary is vague:

    a. measurement error or imprecision (measurement vagueness)
    b. alternative variable combinations in multivariate regions (multivariate vagueness)
    c. boundary changes over time (temporal vagueness)
    d. disagreement about boundary locations (contested vagueness)
    e. fundamentally vague concepts of reality (conceptual vagueness)

These are quite distinct causes of vagueness, with very different philosophical and practical implications (some of which are discussed in Burrough and Frank, 1996). With respect to a core understanding of the region concept (its ontology), two distinctions suggested by this list are most critical. One is between contested vagueness, which allows single persons or groups to identify regions with precise boundaries (just not the same ones) and the other causes, which produce "universally held" vagueness. A second critical distinction in the list is between the last cause, conceptual vagueness, and the first four causes. The first four may always or typically lead to vagueness in practice even though they need not in theory. Conceptual vagueness, in contrast, necessarily leads to vagueness even in theory.

The four types of regions from the taxonomy tend to differ in the extent and nature of their boundary vagueness. All but cognitive boundaries may have measurement vagueness (cognitive regions are not the product of measurement, though they can be measured). Thematic and functional boundaries may have multivariate vagueness (Gray's 1997 proposal for a disaggregate approach to biogeographic regions was inspired by the pervasiveness of multivariate vagueness in that domain). All four boundary types may have temporal or contested vagueness (though only contested vagueness over administrative boundaries tends to create war). And all but administrative boundaries may have conceptual vagueness.

This last point is important insofar as it helps to show the special status of administrative regions. For the other three types, the degree to which places on earth have the property or properties in question typically varies more or less continuously and only a partially arbitrary decision can fully determine boundaries. For example, precisely delimiting an ecosystem region cannot generally be done even in theory. It would be impossible in practice to ever measure all of the natural variables at every location at the same moment in time with perfect accuracy and precision, and if you could, you would not get a completely contiguous region. But fundamentally, the ecosystem region is simply not defined precisely enough as a concept to produce crisp boundaries, whatever the measurement fantasies entertained. Similarly, cognitive and functional regions are typically fundamentally vague, with every crisp representation a fiction to some extent. There is not in principle or in fact a precise boundary around the Bible Belt or the region where people receive the London Times (there must be at least one person in almost every corner of the world, not just the Commonwealth, that receives that paper). None of this is to say there is no transition in reality corresponding to the boundary, only that the transition which really exists is really not sharp. Indeed, the utility of such systems of regions depends in part on their correspondence to states of the real world. Boundary vagueness in no way nullifies this.

Administrative regions are thus unique in being the only one of the four types that typically has the potential for precise boundaries. Couclelis (1992) referred to the crisp boundaries of "regions of social control." Frank (1996) pointed out that precise boundaries come from the "legal objectification" of the earth surface for the purpose of ownership. Smith (1995) discussed the special property of "infinite thinness" that geopolitical boundaries have. However, the uniqueness of administrative boundary precision can be debated. Instances of the other three region types sometimes have thin boundaries, and exceptions to the preciseness of administrative boundaries can be identified (such as that due to contested vagueness). But the special status of administrative regions becomes especially clear when we consider another of their properties, not unrelated to boundary precision: the uniformity of their membership functions. Every place inside of California is completely in California; no places outside of California are in California at all (this holds only for California as an administrative region). Uniform membership functions are generally characteristic of administrative regions, but quite unusual among other region types.

3.1   The fiat/bona fide distinction

Some time ago, Hartshorne (1950) provided an interesting classification of the genesis of administrative boundaries. Among other things, his classification made evident the distinction between administrative boundaries that correspond to real discontinuities in the world and those that do not. Geographers sometimes call this a distinction between "natural" and "artificial" boundaries. In the former case, rivers or mountains or even transitions between cultural group territories provide a basis for the placement of administrative boundaries. In the latter case, boundaries are surveyed in the world or "drawn on a map" in a manner that does not directly correspond to any real transition in the world.

In a series of provocative and influential papers, Smith (e.g. Smith, 1995; Smith and Varzi, 1997) presented this distinction between boundaries that correspond to real discontinuities and boundaries that do not as the basis for a key ontological distinction in the study of regions: *bona fide* vs. *fiat* boundaries. "Bona fide boundaries are boundaries which exist independently of all human cognitive acts—they are a matter of qualitative differentiations or discontinuities in the underlying reality" (1995, p. 476). "Fiat boundaries are boundaries which exist only in virtue of the different sorts of demarcations effected cognitively by human beings" (p. 477). Importantly, at least in his 1995 paper, Smith intends the dichotomy to be "exhaustive and exclusive," though he admits that some boundaries may not fit neatly into one of the two categories (and he allows for mixed cases, as I discuss below).

The distinction between bona fide and fiat boundaries does not map well onto that between administrative and non-administrative regions (which in and of itself does not provide support for or against either dichotomy). In the domain of *geographic* regions, only administrative regions are recognized by Smith to have fiat boundaries. He and Varzi (1997) give as examples of fiat boundaries: "national borders..county- and property-lines..postal districts" (p. 104). But importantly they restrict these examples to: "..those cases where..they lie skew to any qualitative differentiations or spatial discontinuities..in the underlying territory," i.e. fiat boundaries are administrative boundaries not defined by features. Thus, Smith (1995) points out that some administrative boundaries are all fiat, some are mixed bona fide and fiat, and others are entirely bona fide pieces which are "glued together..fiat fashion."

I believe the distinction between bona fide and fiat boundaries is overdrawn, at least with respect to geographic regions. Although physical features often help mark these boundaries, the features are generally not the boundaries, only markers for the boundaries. As I stated above, administrative regions do not exist in the least *as administrative regions* without human intentionality, whether their boundaries are fiat or bona fide. In the southwestern U.S., there is a place where the administrative boundaries of four states come together (Arizona, New Mexico, Colorado, Utah) known as Four Corners. There is a plaque on the ground there that innumerable tourists have gleefully straddled with their four limbs (depending on the time of day you are reading this, it is happening now). But the plaque is not the boundary in the least, and a muscular plaque thief could not modify the administrative boundary in the least. Another example is provided by the American town of Carter Lake, Iowa (discussed in the physical geography text by Christopherson, 1997). It was originally founded within the crook of a meander of the Missouri River, with the center of the river providing a marker for the boundaries of the states of Iowa and Nebraska. When the meander became an oxbow lake, as they are wont to do, the town found itself on the Nebraska side of the new river course. But the town did not become part of Nebraska; the boundary stayed put, leaving a seemingly odd protuberance of Iowa into what is otherwise Nebraska.

On its face, the distinction between fiat and bona fide boundaries appears valid. In the previous section, it turned out to be quite difficult to reduce region types below a dichotomy of administrative and non-administrative regions, but that is not the same as fiat and bona fide. When pushed further, it appears that the fiat/bona-fide distinction is less fundamental. Even Smith has recognized the ultimate lack of distinction between the two. In his 1995 paper, he pointed out that fiat boundaries are in fact always related to bona fide entities. The surveyor establishes fiat boundaries with the use of coordinate systems defined by physical features. In many current contexts that is the earth's shape and rotation, and the observatory marker at Greenwich. In metes-and-bounds systems, it is rocks, tree lines, and rivers. Is there a profound difference between using a river as a border vs. using some number of meters from the river as a border?

4 THE MIND-BODY PROBLEM REARS ITS UGLY HEAD

A dialogue between the Accountant and the Swineherd, over lunch at the Café Earth.

Accountant: I've been reading the new National Geography Standards. They talk about regions, and they say that regions aren't real, they're just mental fictions. Here on page 34, they call them "those human constructs called regions." Could you pass the pepper mill?

Swineherd: That's silly. Anyone who has actually looked around would know that regions are real, they exist. We raise pigs here but they don't raise them up north. This is a pig-raising region, sure enough. Are you saying my pigs aren't real?

A: Your pigs are real enough, I can smell them. But Farmer Jones next door raises chickens, not pigs. And there must be at least one or two farmers up north who have pigs. So the region isn't real, it's just a way of organizing our impressions of the world around us. I could make the region of pigs anywhere else I wanted to.

S: You could but you'd be wrong. Everyone knows this is where most of the pigs are, and if you counted all of the pigs and put that on a map, you would see that there really is a pig-raising region. There are more pigs per hectare here than up north, and that's a fact. Any parsnips left there?

A: Sure, here you go. But your pig region is not full of pigs, and there are more of them in some places than others. How do you decide where to put the boundary? That's not real, it's a subjective decision. What about Scots Grove, where we used to go after school to drink beer? I've never seen that on any map, it was just someplace we all made up. Heck, nobody even remembers who the Scottish guy was.

S: Yeah but we always talked about it like it was real. My mom and dad first met there over 40 years ago. Should I tell them they didn't really meet anywhere, they just thought they did?

A: No no, you've got it confused. They really met, and there really is a physical place where they met. Just calling it Scots Grove is a human invention. I mean if it's a real region, it has to have boundaries. Where are the boundaries?

S: Most people around here know just where they are. As soon as you get over the top of the hill, where you can see the old well, but you're on this side of the swamp, you're there. You're in the Grove. You can "measure" it by asking folks, their answers will tell you.

A: But people just invented that.

S: So? Cars were invented by people, aren't they real?

A: Well only people around these parts know about the Grove. It's not real to anybody else. You can't tell somebody to go there if they aren't from these parts. It's only in our minds, not anybody else's. Are your lamb chops overdone?

S: Not at all, a perfect pink. Again you've got your ontology all messed up. I plan to be rich one day. It's something I think about while I'm slopping my pigs. My plan is real, it exists. I'm not rich yet but I sure as heck am really thinkin' about it.

A: So mental states are real after all. They're not tangible like a pig but just as real.

S: I'll do you one better. How about Goodman County, is that real?

A: Why sure it is, you can see a map with the boundaries of the County on it down at the administration building. I know that's real because my clients have to pay those high property taxes if they live inside the County. Don't tell me something isn't real if you have to pay for it.

S: Now there I will have to differ with you. Hmm, these double-baked new potatoes are delicious. The County is just made up by people, there's nothing real in the world that marks Goodman from McMaster County.

A: Are you kidding? The English River runs right along there, that's where the border is. Have some more broccoli, your hollandaise sauce is getting cold.

S: Have you read the philosopher Smith? He says some boundaries physically exist independently of human cognitive acts. Your river would be an example. He says other boundaries exist only because of human cognition.

A: Yeah, I've read him. He helps to pass the time between tax seasons. I remember at one place, he said that the horizon is a boundary created by cognition. I don't get that, the horizon results from a geometric relationship between a physical earth surface and a physical human body with eyes of a certain sort. That doesn't seem like a cognitive act really. I mean the horizon would still be defined even if the observer was unconscious or dead—she just wouldn't know it!

S: But you know, it's kind of funny. After all Smith's claims in favor of at least two fundamental types of boundaries, he ends up recognizing that even political boundaries involve measurement coordinates based on physical features in the world.

A: As Smith says, it's very difficult to disentangle epistemological from ontological indeterminacy! I don't know what that really means but it sure sounds good. How is your crème brulée?

S: Delightful. At least Smith isn't one of those silly idealists who think all regions are just human conceptions.

A: But are you one of those silly physicists who thinks that all boundaries are in the physical world, as long as you look with a strong enough microscope?

S: Garçon! May we have the check? My friend here won't mind picking it up, since he can just pretend it's not real.

Waiter: How was everything?

S: Everything was really great, except the wine. We don't like our wine with too much oak in it around here.

A: I guess we're really in a region of non-oakies!

Waiter: Mind-body interactionism rules!

S: Say, that reminds me, you heard the one about what time means to a pig?

A: No, but I have to run. Same place next week?

5   CONCLUSION

Would it be possible to do without the concept of regions in geography, particularly when carried out digitally? What alternative is there to the regional organization of geographic information? One is a description or model consisting of continuous functions that cover the entire planet. Another is a dense raster representation in which each small cell contains the values of all the measured variables relating to the entities and processes found in the cell. In either case, individual variables would be left unassociated or independent; they could be described after the fact as correlating spatially in various highly complex fashions. Neither of these approaches is useful for administrative or cognitive regions, however, and the possibility of describing the complex variation of most variables over the entire earth with continuous functions seems remote or impossible. It is true that some global modeling in physical geography, such as climatology, treats the earth as a field (or a fine discrete approximation thereof). Do trained experts think continuously about these phenomena? This is an interesting question that deserves research. My intuition is that relative discontinuities in continuous fields are thought of as vague boundaries by these experts, and that vacant areas in the fields are understood as regions with the presence or absence of some content. Though a computer model may not organize data regionally, displaying it for humans, and thinking and talking about, still falls back on the familiar region.

The importance of regions will continue with the advent of digital geographic information systems and science. Similarly, the enthusiasm in some quarters for a scientific and quantitative approach to geography may modify the role of the region concept, but it does not do away with it. This is true not only because our data processing and analysis methods contain the vestige of pre-digital regionalization, but because regionalization still has great utility. Regionalization is still efficacious and practical for human cognition, no matter the extension of geography to digital representation and analysis in computational systems. Geographic education, when understood in the traditional sense of a concise description of the human and natural earth surface differentiated over space, will always require some form of the region concept (Kostbade, 1968). Visualization of geographic information will always benefit from thoughtful regionalization. Data often comes in regional units, but even when it does not, it is often most effectively communicated via choropleths or isolines that display a regionally organized earth. Attempts to depict only continuous fields on displays will not, in any case, circumvent the human tendency to organize information into discrete units; perceptual grouping and clustering occurs even as continuous fields are being viewed. Furthermore, regionalization is efficacious, even necessary, for the administration of earth space; this echoes Frank et al. (2001), who claimed that administrative regions (they called them *socio-economic regions*) are "necessary for human understanding of space and its administration, despite their shaky ontological foundation" (p. 9).

The conceptual and practical problems of regionalization are considerable; they have long been recognized. It is certainly misleading to speak of one "true" regionalization, and certainly true that the discipline of geography should not content itself with an exclusively regional, as opposed to systematic, approach to understanding its subject matter (Kimble, 1951). But the fact that the regional concept has an imperfect correspondence with measured reality, and the fact that an exclusive focus on regional identification is an inappropriate job for an entire discipline, does not require we dispense with the concept or entertain the fantasy that one day we will be able to do geography without it. An attempt to cleanse regions from geography would be equivalent to throwing out the baby with the bath water. Such an attempt is not required to certify the scientific credibility of the discipline. Regionalization is just an attempt to parsimoniously characterize the earth's surface, to identify the most general truths possible about it. The attempt to identify general truths is highly characteristic of a scientific approach.

ACKNOWLEDGMENTS

I thank Mike Goodchild for his keen mind and equanimity, and the National Center for Geographic Information and Analysis for financial support. Valuable comments by those attending the "The 'I' in GIScience" meeting helped improve the manuscript (though I am confident they would still disagree with various assertions of mine). Special thanks to Violet Gray and David Mark for their regional insights and geographic inspiration.

REFERENCES

Aitken, S. C. (2001). Critically assessing change: Rethinking space, time and scale. In Frank, A. U., Raper, J., and Cheylan, J. P. (Eds), *Life and Motion of Socio-Economic Units*, pp. 189–202. London: Taylor & Francis.

Burrough, P. A. (1996). Natural objects with indeterminate boundaries. In Burrough, P. A. and Frank, A. U. (Eds), *Geographic Objects with Indeterminate Boundaries*, pp. 3–28. London: Taylor & Francis.

Burrough, P. A. and Frank, A. U. (Eds) (1996). *Geographic Objects with Indeterminate Boundaries*. London: Taylor & Francis.

Casati, R. and Varzi, A. C. (1999). *Parts and Places: The Structures of Spatial Representation*. Cambridge, MA: MIT Press.

Couclelis, H. (1992). People manipulate objects (but cultivate fields): Beyond the raster-vector debate in GIS. In Frank, A. U., Campari, I., and Formentini, U. (Eds), *Theories and Methods of Spatio-Temporal Reasoning in Geographic Space*, volume 639 of *Lecture Notes in Computer Science*, pp. 65–77. Berlin: Springer-Verlag.

Fisher, P. (1996). Boolean and fuzzy regions. In Burrough, P. A. and Frank, A. U. (Eds), *Geographic Objects with Indeterminate Boundaries*, pp. 87–94. London: Taylor & Francis.

Forest, B. (2001). Mapping democracy: Racial identity and the quandary of political representation. *Annals of the Association of American Geographers*, 91:143–166.

Frank, A. U. (1996). The prevalence of objects with sharp boundaries in GIS. In Burrough, P. A. and Frank, A. U. (Eds), *Geographic Objects with Indeterminate Boundaries*, pp. 29–40. London: Taylor & Francis.

Frank, A. U., Raper, J., and Cheylan, J. P. (Eds) (2001). *Life and Motion of Socio-Economic Units*. London: Taylor & Francis.

Freksa, C. (1991). Qualitative spatial reasoning. In Mark, D. M. and Frank, A. U. (Eds), *Cognitive and Linguistic Aspects of Geographic Space*, pp. 361–372. Dordrecht, The Netherlands: Kluwer Academic Publishers.

Friedman, A. and Brown, N. R. (2000). Reasoning about geography. *Journal of Experimental Psychology: General*, 129:193–219.

Geography Education Standards Project (1994). *Geography for Life: National Geography Standards*. National Geographic Research & Exploration.

Golledge, R. G. (1992). Do people understand spatial concepts: The case of first-order primitives. In Frank, A. U., Campari, I., and Formentini, U. (Eds), *Theories and Methods of Spatio-Temporal Reasoning in Geographic Space*, volume 639 of *Lecture Notes in Computer Science*, pp. 1–21. Berlin: Springer-Verlag.

Goodchild, M. F. (1992). Analysis. In Abler, R. F., Marcus, M. G., and Olson, J. M. (Eds), *Geography's Inner Worlds: Pervasive Themes in Contemporary American Geography*, pp. 138–162. New Brunswick, NJ: Rutgers University Press.

Gray, M. V. (1997). Classification as an impediment to the reliable and valid use of spatial information: A disaggregate approach. In Hirtle, S. C. and Frank, A. U. (Eds), *Spatial Information Theory: A Theoretical Basis for GIS*, volume 1329 of *Lecture Notes in Computer Science*, pp. 137–149. Berlin: Springer-Verlag.

Hargrove, W. W. and Hoffman, F. M. (1999). Using multivariate clustering to characterize ecoregion boundaries. *Computers in Science and Engineering*, 1:18–25.

Hartshorne, R. (1950). Functional approach to political geography. *Annals of the Association of American Geographers*, 40:95–130.

Hartshorne, R. (1959). *Perspective on the Nature of Geography*. Chicago: Rand McNally & Co.

Hirtle, S. C. and Jonides, J. (1985). Evidence of hierarchies in cognitive maps. *Memory and Cognition*, 13:208–217.

Kaufman, L. (1974). *Sight and Mind: An Introduction to Visual Perception*. New York: Oxford University Press.

Kimble, G. H. T. (1951). The inadequacy of the regional concept. In Stamp, L. D. and Woolridge, S. W. (Eds), *London Essays in Geography*, pp. 151–174. Cambridge, MA: Harvard University Press.

Kostbade, J. T. (1968). The regional concept and geographic education. *The Journal of Geography*, 67:6–12.

Landau, B. and Jackendoff, R. (1993). "What" and "where" in spatial language and spatial cognition. *The Behavioral and Brain Sciences*, 16:217–265.

Linton, D. L. (1951). The delimitation of morphological regions. In Stamp, L. D. and Woolridge, S. W. (Eds), *London Essays in Geography*, pp. 199–217. Cambridge, MA: Harvard University Press.

Maki, R. (1981). Categorization and distance effects with spatial linear orders. *Journal of Experimental Psychology: Human Learning and Memory*, 7:15–32.

Mark, D. M. and Csillag, F. (1989). The nature of boundaries on 'area-class' maps. *Cartographica*, 26:65–78.

Mark, D. M., Skupin, A., and Smith, B. (2001). Features, objects, and other things: Ontological distinctions in the geographic domain. In Montello, D. R. (Ed), *Spatial Information Theory: Foundations of Geographic Information Science*, volume 2205 of *Lecture Notes in Computer Science*, pp. 489–502. Berlin: Springer-Verlag.

Martin, G. J. and James, P. E. (1993). *All Possible Worlds: A History of Geographical Ideas*. New York: John Wiley & Sons, 3rd edition.

McNamara, T. P., Hardy, J. K., and Hirtle, S. C. (1989). Subjective hierarchies in spatial memory. *Journal of Experimental Psychology: Learning, Memory, and Cognition*, 15:211–227.

Meyer, W. B., Gregory, D. B., Turner, B. L., and McDowell, P. F. (1992). The local-global continuum. In Abler, R. F., Marcus, M. G., and Olson, J. M. (Eds), *Geography's Inner Worlds: Pervasive Themes in Contemporary American Geography*, pp. 255–279. New Brunswick, NJ: Rutgers University Press.

Montello, D. R. (2002). Scale, in geography. In Smelser, N. J. and Baltes, P. B. (Eds), *International Encyclopedia of the Social and Behavioral Sciences*, pp. 13501–13504. Oxford: Pergamon Press.

Muehrcke, P. C. and Muehrcke, J. O. (1992). *Map Use: Reading, Analysis, Interpretation*. Madison, WI: JP Publications, 3rd edition.

Richardson, B. C. (1992). Places and regions. In Abler, R. F., Marcus, M. G., and Olson, J. M. (Eds), *Geography's Inner Worlds: Pervasive Themes in Contemporary American Geography*, pp. 27–49. New Brunswick, NJ: Rutgers University Press.

Smith, B. (1995). On drawing lines on a map. In Frank, A. U. and Kuhn, W. (Eds), *Spatial Information Theory: A Theoretical Basis for GIS*, volume 988 of *Lecture Notes in Computer Science*, pp. 475–484. Berlin: Springer-Verlag.

Smith, B. and Varzi, A. C. (1997). Fiat and bona fide boundaries: Towards an ontology of spatially extended objects. In Hirtle, S. C. and Frank, A. U. (Eds), *Spatial Information Theory: A Theoretical Basis for GIS*, volume 1329 of *Lecture Notes in Computer Science*, pp. 103–119. Berlin: Springer-Verlag.

Smith, E. E. and Medin, D. L. (1981). *Categories and Concepts*. Cambridge, MA: Harvard University Press.

Stevens, A. and Coupe, P. (1978). Distortions in judged spatial relations. *Cognitive Psychology*, 10:422–437.

Tajfel, H. and Wilkes, A. L. (1963). Classification and quantitative judgment. *British Journal of Psychology*, 54:101–114.

Talmy, L. (1983). How language structures space. In Pick, H. L. and Acredolo, L. P. (Eds), *Spatial Orientation: Theory, Research, and Application*, pp. 225–282. New York: Plenum Press.

Tversky, B. (2002). Navigating by mind and by body. Paper presented at the International Conference on "Spatial Cognition", DFG Spatial Cognition Priority Program (Evangelische Akademie in Tutzing, Germany).

CHAPTER 10

Neighborhoods and Landmarks

Stephen C. Hirtle
School of Information Sciences
University of Pittsburgh, Pittsburgh, PA, 15260, USA

1 INTRODUCTION

In examining the question of the nature of information for geographic information science (GIScience), multiple levels of information should be considered. Within knowledge management, there is the well-known and often cited ordering of information, which states that with enough processing signals become data, data become information, information becomes knowledge and, eventually with enough processing, knowledge becomes wisdom. Unfortunately, the context of this ordering is rather general. As Coombs (1983) stated in discussing the principle of Pareto optimality, scientists within individual domains have the ability to develop detailed theories, which are more limited in scope, but more powerful in terms of explanatory power. Given the richness of the GIScience domain, it is worth considering more powerful theories of information.

Albert Borgmann (1999) recently addressed the nature of information in a very different way. Borgmann eloquently argued that information over the history of humans has systematically evolved. Early information sources were explicitly tied to nature. Information would emerge by noting the meaning associated with a natural event, such as realizing that birds circling over a certain spot along a river might indicate the presence of fish. Natural information might also occur through the manipulation of the environment, such creating a tower of stones to indicate a path to follow. In either case, natural information is only meaningful in the geographical location where the information occurs. Cultural information, which includes writing and architecture, is more transient in place and time. Writing and literature are obvious examples of cultural information, but Borgmann also includes architecture in this category, as a building tells stories of lives past, as well as directing current activities. Borgmann closes his monograph by positing the technological information takes information one step further. With the advent of technological information, information becomes a reality with content for its own sake. Furthermore, geographic information systems are viewed by Borgmann as the quintessential example of technological information with its ability to reveal otherwise invisible things about physical and social realities of the earth.

In fact, geographic information is a special kind of information that is both tied to location and tied to social processes. That is, geographic information unlike most other forms of information is tied to an inherent reference system of the earth. In addition, geographic thought, unlike geological inquiry, is most often tied to either human behavior or group processes. Recent developments in GIScience have highlighted one, and sometimes both, of these two distinct aspects of geographic information. In this paper, the focus is on the human component and how cognitive processes drive the processing of spatial knowledge.

2 STRUCTURING OF SPACE INTO REGIONS

De Castro (1997), Mark et al. (1999), McNamara (1986), Tversky (1993) and others have written extensively on the structuring of space. It is clear from decades of research that the cognitive representations of space are highly structured. Space tends to be organized into hierarchical or quasi-hierarchical structures. The most common quasi-hierarchical graph structure for modelling cognitive maps is known as a semi-lattice (Alexander, 1965; Hirtle, 1995; Kokla, 2000). Here components of space are structured in overlapping sets or clusters. Items within a cluster are perceived as being closer (Hirtle and Jonides, 1985; McNamara et al., 1989) and clusters are often rotated or translated to align with global coordinate systems (Tversky, 1993)

Figure 1: Eye movements in response to a picture (Source: Yarbus, 1967, p. 180)

Even at the perceptual level structural biases exists. As early as 1967, Yarbus has shown that eye movements across an image are not random. However, it also is interesting to note that they do not follow a regular pattern, such as radar sweeping across the sky. Rather, Yarbus showed that eye movements latch onto specific landmarks or boundaries in the image. Figure 1 shows one such example from Yarbus (1967) of eye movements in response to looking at a photograph.

The process of subitizing and Gestalt groupings leads to another interesting example (Mandler and Shebo, 1982; Trick and Pylyshyn, 1994). If shown a display, such as the one shown in Figure 2a, and asked to count the number of points, most subjects immediately notice 5 points in the upper left and 4 points in the lower left for a total of 9 points. The five points are perceived almost automatically, without the need to count the points, as are the 4 points in the lower left. There is a limit to the number of points that can be perceived as a single unit, typically around five or six. Furthermore, spatial groupings drive perception through Gestalt laws. For example, it is not possible to think of the drawing shown in Figure 2b as a group of 5 o's and 4 x's without great cognitive effort.

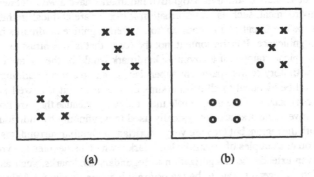

Figure 2: (a) Subitizing and Gestalt principles lead to a grouping of points. (b) An display that is difficult to perceive as 5 o's and 4 x's

Therefore, we see that early perceptual processes can isolate landmarks and boundaries, while Gestalt principles can lead to the formation of natural categories. In the following sections, the emergent properties of landmarks and neighborhoods are described in more detail, followed by a discussion of the how these concepts can be incorporated into geographical information systems.

2.1 Landmarks

Landmarks, as discussed by Golledge (1999), Allen (1999), Presson and Montello (1988), and others, are distinctive elements of environment, which can serve at least two purposes. Landmarks can serve as organizing concepts for the development of a cognitive map or landmarks can serve as navigational tools for wayfinding (Golledge, 1999). As organizing concepts for space, landmarks can represent a cluster of objects or serve as spatial reference points (Sadalla et al., 1980). Landmarks enable one to encode spatial relations between objects and paths, enhancing the development of a cognitive map of a region

(Heth et al., 1997). Alternatively, as navigational aids, landmarks provide critical decision points in navigation, identify important places or regions and allow a traveller to
verify a route (Sorrows and Hirtle, 1999).

Sorrows and Hirtle (1999) defined three categories of landmarks, which apply to both
real environments and virtual environments, such as the World Wide Web. The three
categories are visual landmarks, cognitive landmarks, and structural landmarks. These
categories are not mutually exclusive and often the strongest landmarks are those that are
landmarks in all three senses.

Visual landmarks are structures in the environment, either built or natural, whose
visual characteristics contrast with the surroundings. Examples of visual landmarks can
include obvious referents, such as the Eiffel Tower in Paris and the Washington Monument
in Washington, D.C. However, they can also include much less prominent buildings, such
as a distinctive building at key intersections, which are used in navigation.

Cognitive landmarks are landmarks because of the meaning attached to the structure,
even if it is not visually distinctive. Cognitive landmarks have a well-defined, important
role in the environment, such as an information kiosk, or are atypical in their surroundings, such as the apartment of a famous author. Often cognitive landmarks have cultural
or historical significance. It is the content, not the form, that is in contrast to the surroundings. A prototypical example of a cognitive landmark would be the room of the resident
advisor in a dormitory or the apartment supervisor in an apartment building. The room
might very well be identical to all other rooms on the floor, but it is well known to the
floor's residents because of the unique role that it serves. Because they are not distinctive
visually cognitive landmarks are not typically used in wayfinding by individuals who are
unfamiliar with the region, but they are very important in familiar surroundings. Portugali
(2000) gives other examples of cognitive landmarks, which he denotes as symbolic landmarks. He even extends the categorization to legendary landmarks, such as the Verona
balcony, which by legend is said to be the original location of famous dialogue between
Romeo and Juliet. The balcony is one of the most photographed, yet visually it is rather
plain, not unlike hundreds of other balconies in the region. Furthermore, the famed dialogue is only legend and most likely never occurred at this spot. Yet for residents and
tourists alike it has become a central landmark for the city.

Structural landmarks are those in which the role or location in the space is notable.
Structural landmarks are highly accessible and have prominent location. An example of a
structural landmark would be a central plaza or train station from which numerous other
locations are reachable. Structural landmarks are clearly noted in a graphical representation of a city. This notion of "landmarkedness" has been incorporated in to the identification of landmarks in hypertext, as well. For example, Mukherjea and Hara (1997), define
the structural importance of the node in terms of how many links lead from a page and
how many links lead to a page, among other measures. This graph-theoretic calculation
could easily be adopted in the physical environment (Sorrows and Hirtle, 1999).

2.2 Neighborhoods

Neighborhoods are an important organizing construct and are discussed in detail by Montello (this volume) and Galton (this volume). In the context of this article, neighborhoods
have two important properties. They are hierarchically structured and they may be inde-

terminate or vague with the respect to boundaries. On the basis of these two properties, a wide a range of terminology can be developed. Any set of neighborhood terms can be plotted in at least two dimensions, as shown in Figure 3. In Figure 3, examples of spatial area terms are plotted in dimensions of informal to formal terms and small to large spaces. Plot and tract are legal terms for a relatively small parcel of land, which is shown in the lower right corner of the graph, whereas urban area is non-legal term for describing a city, shown in upper left corner. Regional differences and legal regulations determine terminology used at each level by different parts of the English-speaking world.

The hierarchical structure that is inherent in spatial terminology can be seen through chains of terms, such as yard → lot → block → community → city → region. Yard and lot refer to property owned by a single individual, while community might refer to part of a city with similar housing and shared shopping districts (Hirtle, 1995; Lloyd et al., 1996; Tversky and Hemenway, 1983).

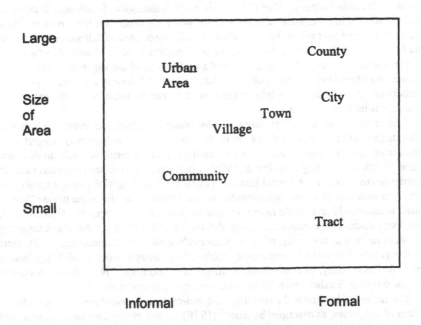

Figure 3: Plotting neighborhoods and regions as function of size and formality of terminology

2.3 Relationship between landmarks and neighborhoods

Numerous authors have posited that cognitive maps are organized around landmarks and neighborhoods, among other structural objects, such as edges and paths (Allen et al., 1979; Golledge, 1999; Couclelis et al., 1987; Lynch, 1960; Passini, 1984). There is an interesting relationship between neighborhoods and landmarks, which in turn has strong implications for the usability of geographic information systems. Landmarks, which refer to points in space, are complementary to neighborhoods, which consist of small regions in a space. A neighborhood can be defined as the region surrounding one or more landmarks and a landmark can be defined as the most prominent building in a neighborhood. There are even geometric tools, such as Voronoï diagrams and Delaunay triangulation, which allows one to establish equivalencies between points and regions (Edwards et al., 1996; Gold, 1992). However, despite the correspondence between landmarks and neighborhoods, there are also critical distinctions in how and when each are used in discourses concerning spatial referents.

2.4   Fuzziness of landmarks and neighborhoods

What if something is not quite a landmark or not quite a neighborhood? How is fuzziness captured in the concepts? The fuzziness of a landmark tends to be tied to the conceptual notion of "landmarkedness." That is, there are some buildings or locations that are more likely to act as a landmark for most individuals, while are others are more personal landmarks or contain fewer of the characteristics typically associated with landmarks (Sorrows and Hirtle, 1999). In contrast, the fuzziness of a neighborhood is often tied to the extent or size of the region, in that the boundaries of a neighborhood are vague or indeterminate.

Smith and Mark (1998) have made a similar distinction between a mountain peak and a mountain. A peak is a well-defined location at a specific point, while the extent of a mountain is inherently vague.

Indeterminate boundaries are suggested by many neighborhood terms, such as outlying district, urban area, or suburia. In other cases, the terminology may suggest crisp boundaries, often by legal mandate, but in reality the use of term may still imply a gradation of values (Burrough and Frank, 1996). For example, there are numerous cases of "border towns" that have a special free-trade agreement or a right-of-passage agreement, even if in violation of a strict interpretation of the national law. Even the phrase "border town" is rather odd and would have no meaning under a strict hierarchy. However, it is a perfectly reasonable concept under a graded membership function. Another example is the oft cited "not in my backyard" phenomena (Couclelis and Monmonier, 1995). Here conflict is caused by an undesirable entity, such as town dump or county jail, being located near a desirable entity, such as school or affluent neighborhood. "Near" in this context is not just physical distance and reinforces the notion that boundaries are graded.

The notion of hierarchical structuring and indeterminate boundaries is similar to the notion of categories, as described by Rosch (1978). In describing categories, Rosch used the term vertical structure to refer to subordinate, basic and superordinate category terms. An example of each of these categories might be electric guitar, guitar, and musical instrument. Whether or not space has a similar level as a basic level category is an open question (Smith and Mark, 1998; Lloyd et al., 1996). The horizontal structure of a category is given by the graded membership with prototypical exemplars in the center of a category. Under Rosch's theory 'robin' would be a prototypical exemplar of the category 'bird', whereas 'penguin' would be an atypical exemplar. Switching to a spatial context, one might find buildings that are clearly within downtown, others that are on clearly not part of downtown and others that are on the fringe of downtown. That is, the borders of a given neighborhood are often vague and indeterminate (Smith and Mark, 1998).

The term 'landmark' is not nearly so rich in providing the kind of Roschian vertical and horizontal structure, which is found with the term 'neighborhood'. Instead, one often refers to landmarks with name of the object (e.g., turn left at the Exxon station) or name of superset (e.g., turn left at the gas station). There is no equivalent to the generic neighborhood terminology of "district" or "lot," such as "big square building."

2.5 Creating structure

Perhaps of most interest is the ability for humans to create structure where there is none. As one demonstration of how this phenomenon might occur, McNamara et al. (1989) showed subjects a room full of objects or a map of objects with no apparent neighborhood structure. Subjects were required to memorize the location of all the objects, which numbered around 30 between the two studies. McNamara et al. showed that subjects naturally performed the task by creating clusters of objects in nearby locations. Furthermore, their recall of locations was biased by the clusters, such that within-cluster pairs were judged to be closer than equidistant between-cluster pairs.

In related research, Portugali (1999) has shown that, even in the absence of high-order plans, spatial systems often appear to develop organizational structures. While this is true of many complex systems, it is particularly striking with regard to cities. In one empirical study, Portugali (1996) had subjects place blocks, one at time, on a flat surface to create a miniature town. No instructions were given to the subjects and no constraints were put on the placement of the blocks. However, in all simulations, a structure emerged, which can be described by a theory of inter-representation networks and self-organizing systems.

3 IMPLICATIONS

The most important issue is to consider what the implications of this analysis are as we build geographic information systems. The implications are necessarily complex and can be separated into two distinct scenarios. It is argued that many modeling and simulation systems must operate in a cognitively naïve manner. The inclusion of cognitive biases is not relevant and if included would result in sub-optimal performance of the models.

However, other systems, such as geographical decision support systems, must recognize biases to generate plausible explanations for seemingly incorrect solutions. Consider a typical route finding system, such as http://www.mapquest.com. Table 1 gives a simple set of directions. Such directions are difficult to follow because all segments, be it an alley or major interstate, are described with equal weight. The entire set of instructions could be replaced with "Take I-270 North." The instructions are also confusing as there is no mention of neighborhoods or landmarks, which are often given in route descriptions generated by individuals.

Structural knowledge is also critical for providing users with validation of decisions. For example, directions, which appear to be overly circuitous, will be ignored, unless justification can be given. It has become common in artificial intelligence to present near misses, in addition to the preferred answer, as a means of explanation. Intelligent agents also use a similar process. An agent, when asked to schedule a flight from Boston to Pittsburgh sometime after 6pm, might respond that there is a 1.5 hour flight at 5:30pm and a 2 hour flight at 8pm. The first flight fails the strict criterion that was given, but, in fact, might be preferable to the traveller. Likewise, an intelligent geographic information agent should provide near misses, particularly if the optimal solution appears to be circuitous, which can be formally defined as a deviation form in lattice of neighborhoods in the cognitive map.

14.	Take the I-270 NORTH exit\	0.13 mi
15.	Merge onto I-270 N.	3.46mi
16.	Take I-270 LOCAL N towards MONTROSE RD.	0.2mi
17.	Merge onto I-270 LOCAL N.	3.47mi
18.	Take I-270 NORTH RAMP towards FREDERICK.	0.18mi
19.	Merge onto I-270 N.	0.76mi
20.	Take the exit\	0.18mi
21.	Merge onto I-270 LOCAL N.	3.61mi
22.	Take I-270 N.	20mi

Table 1: Partial set of directions from www.mapquest.com

3.1 How might these structures be included?

It is argued that the general usability of systems will improve through imposing higher order constructs on geographical information and through the construction cognitively driven ontologies. As Raskin (2000) argued in recent book, *The Humane Interface*, interfaces should be responsive to human needs. Object oriented programming techniques encourage software developers to make meaningful units of analysis explicit. Gero and Tversky (1999) describe numerous applications where visual and spatial reasoning in design is facilitated greatly by modeling the design process using cognitively plausible models.

However, perhaps the most promising approach is to incorporate the use of ontologies (Guarino, 2000; Guarino and Giaretta, 1995; Winter, 2000). At the National Science Foundation Research Planning Workshop on Cognition and Spatial Reasoning, (Epstein et al., 1997) the importance of spatial cognition as a fundamental process for interaction with the environment was delineated. In particular, the workshop focused on the human-machine connection. Among several agenda items, the report argued for additional research support for the ontological foundations of spatial reasoning with application to both computer-aided design (CAD/CAM) and geographic information systems

(GIS). This theme was reiterated in a Workshop on Geographic Information Science and Geospatial Activities at NSF (Mark, 1999), where the call for basic research in geographic information science (GIScience), including the ontological questions regarding the nature of space. More recently, the European Science Foundation sponsored a EuroConference on Ontology and Epistemology for Spatial Data Standards in La Londe-les-Maures, September 22-27, 2000 to bridge the gap between research on ontology with research in representational models (Winter, 2000).

The bridge between ontology and representational models with respect to spatial cognition can lead directly to advances in interface design for geographic information systems. In particular, navigation systems will have the potential to be enriched with cognitively meaningful information that will improve the usability of geographic information systems. As one example, consider the simple problem of trying to find a hotel near a certain tourist district. As with the previous route-finding example, the lack of understanding of the human conception of neighborhoods, as opposed to a legal or formal conception, can result in strange or obtuse behavior.

To understand cognitive maps that individuals have constructed of space and, at the same time, to build usable spatial information systems, it is advantageous to develop an ontology of geographic concepts (Mark et al., 2000; Smith and Mark, 1998). In particular, the ontology of neighborhoods and landmarks will provide a central part of a theory of cognitive maps and human wayfinding. In this way, the nature of the concepts of landmark and neighborhood is delineated as a foundation for building cognitively plausible ontology of spatial information. The approach taken melds research spatial cognition (Mark, 1999), with theoretical advances in ontology for GIScience (Winter, 2000).

4 POSSIBLE EXTENSIONS

Material presented in the previous sections suggest features to include in an ontology of neighborhoods and landmarks, but clearly much work needs to be done to create the formal structure needed to employ these ideas in working systems. Traditional part-whole relationships will not adequately represent the complexity of these concepts. Guarino (2000) has described this problem as "is-a overloading," which results in a reduction of the sense of a term or overgeneralization. Kokla (2000) has taken the approach of the generation of a multi-scale, multi-context database-model generalization, which can account for the unique nature of geographical representations, using Formal Concept Theory (Sowa, 2000). This approach can represent overlapping categories by the generation of a lattice, which includes the original categories plus artificial constructed members, which were not originally specified by the cognitive model, for mathematical completeness.

Applications where a neighborhood ontology will improve usability can be explored. These include typical wayfinding systems, as might be found on Mapquest.com or other commercial servers. Instructions on such systems today are very sparse in terms of indicating structural divisions. From an ontological perspective, it appears that it would be beneficial to include additional information about physical features of the environment (such as, traffic lights or elevation), neighborhoods and landmarks. None of these components are common in automated wayfinding systems. Landmarks, as point objects, are easier to incorporate into current systems. In a similar vein, Mark (2000) has argued, a mountain peak is well-defined and is a common cartographic feature, but a mountain, as a region, is inherently vague and is absent as a cartographic feature from virtually all maps. However, there is some empirical work, which suggests how to incorporate regions into route directions (Egenhofer and Mark, 1995). The benefits, or lack thereof, of including neighborhood and landmark information should be measured using empirical studies.

As a final example, consider the example of a low-income resident attempting to locate a job, day care facility, and apartment, all reachable by public transportation. By ordering options, not by distance, but by neighborhood, could greatly facilitate communication of the possible locations. Furthermore, the vector of bus lines would create the primary structure for visualization to the client, as opposed to a simple list of options or a distribution around central points, as one might present for other spatial problems.

5 CONCLUSIONS

From low-level perceptual processes to high-level cognition, space is not processed as a continuous field. Individuals create structure by dividing space into regions, creating artificial boundaries, and identifying landmarks. This phenomenon is so pervasive that even in situations were no higher order structure exists, individuals will manufacture false structures, which in turn will induce biases and errors in judgment.

In this paper, the ontological foundations for landmarks and neighborhoods are explored. Landmarks and neighborhoods are partially complementary terms for organizing space. Landmarks are identified as objects in space and labeled by the name of the object. Neighborhoods are identified as regions with and graded membership and indeterminate boundaries. There is a large vocabulary bridging informal to formal terminology that can denote regions, where each term has implications for the size and connectedness to other units.

The approach taken melds research spatial cognition (Mark, 1999), with theoretical advances in ontology for GIScience (Winter, 2000). Finally, it is argued that wayfinding systems will be lacking in usability, unless both landmarks and neighborhoods are systematically included as part of route descriptions. The overall usability of GIS will be dependent on the inclusion of cognitively plausible structures within the information system (Edwards, 1997; Hirtle, 1998; Mennis et al., 2000).

REFERENCES

Alexander, C. (1965). A city is not a tree. *Architectural Forum*, pp. 58–62.
Allen, G. L. (1999). Spatial abilities, cognitive maps, and wayfinding: Bases for individual differences in spatial cognition and behavior. In Golledge, R. G. (Ed), *Wayfinding behavior: Cognitive mapping and other spatial processes*, pp. 46–80. Baltimore, MD: Johns Hopkins Press.
Allen, G. L., Kirasic, K. C., Siegel, A. W., and Herman, J. F. (1979). Developmental issues in cognitive mapping: The selection and utilization of environmental landmarks. *Child Development*, 56:1062–1070.
Borgmann, A. (1999). *Holding on to reality: The nature of information at the turn of the millennium*. Chicago: University of Chicago Press.

Burrough, P. A. and Frank, A. U. (Eds) (1996). *Geographic Objects with Indeterminate Boundaries*. London: Taylor & Francis.

Coombs, C. H. (1983). *Psychology and mathematics: An essay on theory*. Ann Arbor: University of Michigan Press.

Couclelis, H., Golledge, R. G., Gale, N., and Tobler, W. (1987). Exploring the anchor-point hypothesis of spatial cognition. *Journal of Environmental Psychology*, 7:99–122.

Couclelis, H. and Monmonier, M. (1995). Using SUSS to resolve NIMBY: How spatial understanding support systems can help with the 'Not In My Back Yard' syndrome. *Geographical Systems*, 2(2):83–101.

de Castro, C. (1997). *La geografía en la vida cotidiana*. Barcelona: Ediciones del Serbal.

Edwards, G. (1997). Geocognostics: A new framework for spatial information theory. In Hirtle, S. C. and Frank, A. U. (Eds), *Spatial information theory: A theoretical basis for GIS*, volume 1329 of *Lecture Notes in Computer Science*, pp. 455–472. Berlin: Springer-Verlag.

Edwards, G., Ligozat, G., Gryl, A., Fraczak, L., Moulin, B., and Gold, C. (1996). A Voronoï-based pivot representation of spatial concepts and its application to route descriptions expressed in natural language. In *Proceedings of the 7th International Symposium on Spatial Data Handling*, Delft, The Netherlands.

Egenhofer, M. J. and Mark, D. M. (1995). Modeling conceptual neighborhoods of topological relations. *International Journal of Geographical Information Science*, 9:555–565.

Epstein, S. L., Gelfand, J. J., and Marefat, M. M. (1997). Report on the NSF research planning workshop on cognition and spatial reasoning: The human-machine connection. http://www.princeton.edu/~jig/nsf_report.html.

Galton, A. (2002). On the ontological status of boundaries. In Duckham, M., Goodchild, M. F., and Worboys, M. F. (Eds), *Foundations in Geographic Information Science*, pp. 151–171. London: Taylor & Francis.

Gero, J. S. and Tversky, B. (Eds) (1999). *Visual and spatial reasoning in design*. Sydney, Australia: Key Centre of Design Computing and Cognition.

Gold, C. M. (1992). The meaning of "neighbour". In Frank, A. U., Campari, I., and Formentini, U. (Eds), *Theories and Methods of Spatio-Temporal Reasoning in Geographic Space*, volume 639 of *Lecture Notes in Computer Science*, pp. 220–235. Berlin: Springer-Verlag.

Golledge, R. G. (1999). Human wayfinding and cognitive maps. In Golledge, R. G. (Ed), *Wayfinding behavior: Cognitive mapping and other spatial processes*, pp. 5–45. Baltimore, MD: Johns Hopkins Press.

Guarino, N. (2000). Towards a methodology for ontology design: Identity, unity, and taxonomic constraints. In Winter, S. (Ed), *Geographical domain and geographical information systems, EuroConference on ontology and epistemology for spatial data standards*, La Londe-les-Maures, France. http://www.esf.org/euresco.

Guarino, N. and Giaretta, P. (1995). Ontologies and knowledge bases: Toward a terminological clarification. In Mars, N. J. I. (Ed), *Towards very large knowledge bases*, pp. 25–32. Amsterdam: IOS Press.

Heth, C. D., Cornell, E. H., and Alberts, D. M. (1997). Differential use of landmarks by 8- and 12- year-old children during route reversal navigation. *Journal of Environmental*

Psychology, 17:199–213.

Hirtle, S. C. (1995). Representational structures for cognitive space: Trees, ordered trees, and semi-lattices. In Frank, A. U. and Kuhn, W. (Eds), *Spatial Information Theory: A Theoretical Basis for GIS*, volume 988 of *Lecture Notes in Computer Science*, pp. 327–340. Berlin: Springer-Verlag.

Hirtle, S. C. (1998). The cognitive atlas: Using a GIS as a metaphor for memory. In Egenhofer, M. J. and Golledge, R. G. (Eds), *Spatial and temporal reasoning in geographic information systems*, pp. 263–271. New York: Oxford.

Hirtle, S. C. and Jonides, J. (1985). Hierarchies in cognitive maps. *Memory and Cognition*, 13(3):208–217.

Kokla, M. (2000). Concept lattices as a formal method for the integration of geospatial ontologies. In Winter, S. (Ed), *Geographical domain and geographical information systems, EuroConference on ontology and epistemology for spatial data standards*, La Londe-les-Maures, France. http://www.esf.org/euresco.

Lloyd, R., Patton, D., and Cammack, R. (1996). Basic-level categories. *Professional Geographer*, 48:181–194.

Lynch, K. (1960). *The Image of the City*. Cambridge, MA: MIT Press.

Mandler, G. and Shebo, B. J. (1982). Subitizing: An analysis of its component processes. *Journal of Experimental Psychology: General*, 111:1–22.

Mark, D. M. (Ed) (1999). *Geographic information science: Critical issues in an emerging cross-disciplinary research domain*, NSF Workshop Report. http://www.geog.buffalo.edu/ncgia/workshopreport.html.

Mark, D. M. (2000). Ontology of geographical object categories. In Winter, S. (Ed), *Geographical domain and geographical information systems, EuroConference on ontology and epistemology for spatial data standards*, La Londe-les-Maures, France. http://www.esf.org/euresco.

Mark, D. M., Egenhofer, M., Hirtle, S. C., and Smith, B. (2000). Ontological foundations for geographic information science. UCGIS Emerging Research Themes. Available from http://www.ucgis.org/.

Mark, D. M., Freksa, C., Hirtle, S. C., Lloyd, R., and Tversky, B. (1999). Cognitive models of geographical space. *International Journal of Geographical Information Science*, 13:747–774.

McNamara, T. P. (1986). Mental representation in spatial relations. *Cognitive Psychology*, 18:87–121.

McNamara, T. P., Hardy, J. K., and Hirtle, S. C. (1989). Subjective hierarchies in spatial memory. *Journal of Experimental Psychology: Learning, Memory, and Cognition*, 15:211–227.

Mennis, J. L., Peuquet, D. J., and Qian, L. (2000). A conceptual framework for incorporating cognitive principles into geographical database representation. *International Journal of Geographical Information Science*, 14:501–520.

Montello, D. R. (2002). Regions in geography: Process and content. In Duckham, M., Goodchild, M. F., and Worboys, M. F. (Eds), *Foundations in Geographic Information Science*, pp. 173–189. London: Taylor & Francis.

Mukherjea, S. and Hara, Y. (1997). Focus+context views of world-wide web nodes. In *Hypertext '97: The Eighth ACM Conference on Hypertext*, pp. 187–196, New York,

NY. ACM Press.

Passini, R. (1984). *Wayfinding in architecture.* New York: Van Nostrand Rheinhold.

Portugali, J. (1996). Inter-representation networks and cognitive mapping. In Portugali, J. (Ed), *The construction of cognitive maps,* pp. 11–43. Dordrecht: Kluwer.

Portugali, J. (1999). *Self-organization and the city.* Berlin: Springer.

Portugali, J. (2000). The I in GIS. GIScience 2000, Savannah, GA.

Presson, C. C. and Montello, D. R. (1988). Points of reference in spatial cognition: Stalking the elusive landmark. *British Journal of Developmental Psychology,* 6:378–381.

Raskin, J. (2000). *The humane interface: New directions for designing interactive systems.* Boston: Addison-Wesley.

Rosch, E. (1978). Principles of categorization. In Rosch, E. and Lloyd, B. B. (Eds), *Cognition and Categorization,* pp. 27–48. Hillsdale, NJ: Erlbaum.

Sadalla, E. K., Burroughs, W. J., and Staplin, L. J. (1980). Reference points in spatial cognition. *Journal of Experimental Psychology: Human Learning and Memory,* 5:516–528.

Smith, B. and Mark, D. M. (1998). Ontology and geographic kinds. In *International Symposium on Spatial Data Handling (SDH'98),* Vancouver, Canada.

Sorrows, M. E. and Hirtle, S. C. (1999). The nature of landmarks for real and electronic spaces. In Freksa, C. and Mark, D. M. (Eds), *Spatial information theory: Cognitive and computational foundations of geographic information science,* pp. 37–50. Heidelberg: Springer-Verlag.

Sowa, J. F. (2000). *Knowledge representation: Logical, philosophical, and computational foundations.* Pacific Grove: Brooks Cole.

Trick, L. M. and Pylyshyn, Z. W. (1994). Why are small and large numbers enumerated differently? A limited-capacity preattentive stage in vision. *Psychological Review,* 101:80–102.

Tversky, B. (1993). Cognitive maps, cognitive collages, and spatial mental models. In Frank, A. U. and Campari, I. (Eds), *Spatial Information Theory: A Theoretical Basis for GIS,* volume 716 of *Lecture Notes in Computer Science,* pp. 14–24. Berlin: Springer-Verlag.

Tversky, B. and Hemenway, K. (1983). Categories of environmental scenes. *Cognitive Psychology,* 113(169–193).

Winter, S. (Ed) (2000). *Geographical domain and geographical information systems, EuroConference on ontology and epistemology for spatial data standards,* La Londe-les-Maures, France. http://www.esf.org/euresco.

Yarbus, A. L. (1967). *Eye movements and vision.* New York: Plenum Press.

CHAPTER 11

Geographical Terminology Servers—Closing the Semantic Divide

Christopher B. Jones
Department of Computer Science
Cardiff University, Cardiff, CF24 3XF, UK

Harith Alani
Electronics and Computer Science Department
University of Southampton, Southampton, SO17 1BJ, UK

Douglas Tudhope
School of Computing, University of Glamorgan
Pontypridd, CF37 1DL, UK

1 INTRODUCTION

Most aspects of human experience may be regarded as having a geographical dimension. Thus everything that we do usually takes place somewhere in the vicinity of the Earth's surface and communication between people frequently requires that we refer to particular places. A consequence of this is that many types of information either have, or else need to be given, some explicit geographical context. The last couple of decades has seen great advances in the development of technology referred to as geographical information systems (GIS). In practice these systems are typically concerned with handling digital maps in which location is recorded primarily by geographical (latitude and longitude) or map grid coordinates. For the most part, individual GIS are domain specific, often project-based, serving the needs of organisations that have traditionally relied upon map-based recording of information.

Undoubtedly GIS have made significant contributions to improving the information retrieval and analysis capabilities of these organisations. When viewed with regard to the need for public access to geographically referenced information, the contribution has been less significant. For those with access to the Internet, one of the commonest methods of seeking information is to employ a search engine within a web browser. When a geographical term, such as a place name, is typed in, it is usually treated the same as any other term or phrase, with the result that documents are retrieved if there is an exact match with the whole or some part of the query phrase. The consequence is that we

205

will often fail to find information that we are interested in, because it has been described by geographical terminology different from that of the query, even though it refers to a similar location. This may happen due to the hierarchical nature of geographic space, so that a particular place may have sub-parts or super-parts referred to by different names. In fact there are multiple overlapping hierarchies of geographical place, varying according to political, topographic and cultural perspectives. Equally there may be places that are close to the specified place and hence potentially of interest, as well as place names that may differ due to historical change or to language.

Ideally, when we use a place name to refer to some location, we should be able to retrieve information about places that are equivalent or nearby and rank the results according to their relevance to the query. When we specify a term or phrase referring to the thing of interest we should be able to find things that have equivalent or similar descriptors. At present web search engines are weak in handling all types of terminology where there is a need for imprecise matching between query and target. These shortcomings are widely recognised and have led for example to the development of XML (extensible markup language) vocabularies that can be used to tag data with terms that clarify meaning, as well as improved levels of intelligence in the search engine itself (see for example Guarino et al., 1999). In this paper we are concerned with improving the level of intelligence of information retrieval tools with regard to geographical terminology.

In conventional GIS the most common way of accessing information by location is to point to somewhere on a map or to specify coordinates explicitly. Frequently spatial objects may also have their name as an attribute that can be used for search, but the associated query procedure is normally based on precise match. Some GIS include a simple gazetteer that allows the user to specify a place name that is then used automatically to specify a map coordinate for purposes of coordinate-based search. The importance of place names as a way of allowing users to search for computer-based information was recognised at the Getty Institute in the mid 1990s and led to the development of the Thesaurus of Geographic Names (TGN) (Harpring, 1997). The TGN is hierarchically structured and hence allows for the possibility of expanding a place name query term by finding its contained places and the parent places. It also maintains different versions of the same name, along with their associated dates, and the geographical coordinates of a representative point location, i.e. a *centroid*. Places in the TGN are either geopolitical or topographic and are associated with place types using terms taken from the Art and Architecture Thesaurus (AAT). In parallel with the TGN, various other gazetteers and place name lists have been produced on regional and international levels. In association with the Alexandria Digital Library a gazetteer metadata standard has been developed in which all places are associated with a coordinate-based spatial footprint (such as a minimum bounding rectangle), while other relationships such as of administrative hierarchy may be recorded but are not mandatory (Hill et al., 1999).

The introduction of gazetteers and the TGN has gone some way to addressing the requirement for access to information by place name, but there has been very little research to investigate how they may be exploited automatically and indeed what sort of geographical information they should record to maximise their utility in information retrieval. The problem of providing intelligent support for geographically referenced query on the web, as well as within specialised GIS, is a challenging one (Walker et al., 1992;

Larson, 1996; Jones et al., 1996; Beard and Sharma, 1997; Moss et al., 1998; McCurley, 2001). In principle, for web search it is desirable to be able to recognise place names at all levels of generalisation for anywhere on Earth, find equivalent co-located places and nearby places and rank the results. Many questions arise with regard to the information to be stored: what types of spatial relationship between stored places; how much coordinate-based data; how can imprecise regions be represented; what types of information characterise places from cultural and environmental perspectives? Decisions about what should be stored affect the capability of relevance ranking procedures.

In this paper we address some of these issues and we describe an experimental cultural information system that integrates geographical and thematic thesauri in the context of a semantic modelling system. In section 2 we elaborate on the subject of defining place for purposes of information retrieval, before summarising in section 3 the potential contribution of previous work on thesauri for encoding semantic relations between terminology. The metadata schema of OASIS are presented in section 4 and the subject of semantic closeness measures is considered in section 5 in which we describe techniques for ranking spatial and non-spatial information employed in the experimental system. Concluding remarks and some issues for further research are presented in section 6.

2 ENCODING PLACES FOR INFORMATION RETRIEVAL

In general when place names are specified in a query to retrieve information, the place may serve to specify location, either by itself, or as part of a spatial expression, or it might be used for comparative purposes to find similar types of place. Here we focus on the former role of place as locator. Thus we assume that a query searches for "something" geographically located "somewhere," where the "somewhere" may include a place name. When modelling place for information retrieval, we need to identify characteristics that will assist in expanding the set of query terms to include co-located places (including places with different names but similar spatial extent) and neighbouring or nearby places, and in ranking the results with respect to the query terms. It may be noted that the final ranking should take account both of the phenomena of interest and the geographic location. We consider modelling non-locational concepts in a subsequent section.

Much has been written about the nature of geographic place (e.g. Relph, 1977; Tuan, 1977; Gould and White, 1986; Johnson, 1991; Curry, 1996; Jordan et al., 1998). It is one of several "basic concepts" in geography (Couclelis, 1992) alongside location, region and space, of which space may be regarded as the most fundamental. Thus space may be considered the substrate within which locations, regions and places are defined. We are concerned particularly with place here because it is associated with names for specific parts of space and hence allows us to refer to location in natural language, as opposed to the typically more formalised expressions of location in terms of coordinates and spatial objects. An issue often raised is that of the human characterisation of place, the fact that places materialise in response to events and experiences and hence are essentially a human construction. Examples of distinguishing properties of a place are the name, the categories that reflect its physical or social features, familiar landmarks that become symbolic, activities, personal experiences and opportunities.

In building an ontology of place to support public information retrieval it may be regarded as a priority to store information that is generic in the sense of not being specific to a few individuals. In this regard experiential aspects of place that are personal in nature may be of low priority. From a pragmatic standpoint, assuming that we may need globally extensive coverage, it may be important to select characteristics that are relatively easily obtainable. This may result in a view of place that reflects to some extent that of Johnson (1991), who adopts for place some of the concepts of Paasi concerned with the institutionalization of regions. The reason for this bias is that existing sources of lists of place names such as those of national mapping agencies typically confine their non-geometric attributes to those of administrative authority (and hence typically a containing region) and perhaps the size of the population. The distinction between regions and places may be of significance for purposes of information retrieval in that regions relate to a partition, often hierarchical, of some parts of space. In doing so they provide a representation that can be exploited for query expansion. In so far as regions label parts of space with commonly used names, we regard them as a type of place. The only significant distinction for our purposes is the fact that regions often form systematic partitions and hierarchies while other places could be isolated, while still referenced in some way to regions.

2.1 Identity

As we are concerned here with specific instances of places, it is essential to maintain the name or names that are typically used to refer to the place. Names may be formal administrative terms, that typically will correspond to a precise boundary such as that of a city or parish, and informal terms that reflect common means of referring to places that may be fairly precise in extent, such as a building, or be imprecise such as a mountain range. Place names change over time and in doing so may come to differ somewhat in exactly what territory or phenomenon they refer to. Certainly knowledge of the temporal extent of a place name may be important when searching for information that itself may be temporally specific. Names may also differ simply due to differences in language. A single name is sometimes used to refer to different places which means that a unique identifier must be found or created. The need for explicit unique identification may vary according to the nature of associated data that are stored. If places are always linked to parent regions, or if a geographical coordinate is stored, then the presence of these attributes may serve to obtain uniqueness.

2.2 Spatial data

Query expansion with respect to location can be supported using coordinate-based methods of conventional GIS in which a search is expanded with increasing Euclidean distance from the query object. Standard GIS methods assume the presence of point, line and polygon spatial objects defined by coordinates. If a model of place were to be maintained for the entire globe as might be required for general web browser querying, then the amount of coordinate data required to represent both the smallest and the largest places would be massive, and on first impression impracticable. There is a motivation therefore for a parsimonious spatial model that encompasses much of geographic space but in a way that minimises the amount of stored information. An alternative to dependency upon coordinates is to encode qualitative relations, such as those of containment/inclusion, overlap and contiguity, all of which can be derived from vector map data, which need not subsequently be stored as part of the model. This would then facilitate query expansion to contained and containing places as well as to neighbouring places. Assuming the presence of multiple (overlapping) regional hierarchies, then if a place was to be registered in the ontology, its containment and overlap relations to all hierarchies could be determined and recorded. An advantage of encoding and exploiting qualitative spatial relations is that it enables some historical places to be recorded for which documentary evidence provides regional containment information in the absence of a cartographic representation. In section 5 we discuss the issue of measuring similarity of place using hierarchical relationships.

It should be noted that contiguity relationships can only easily be derived from maps based on polygonal partitions of space (as in administrative regions) and hence will lead directly to query expansion only within the map regions. For purposes of determining proximal and directional relations between isolated places, it may still be very useful to employ some coordinate data. At the least this could be a single representative point or centroid, as in a simple gazetteer. Storage of centroids facilitates ranking of places with respect to Euclidean distance and the determination of nearest neighbouring places using Voronoi diagram or Delaunay triangulation methods (Aurenhammer, 1991). Distances calculated between places based on centroids will of course be approximations as they take no account of the location of the boundary. It is however possible to estimate the locations of boundaries of places given data on contained and neighbouring places for which centroids are available (Alani et al., 2001).

2.3 Accessibility

When considering "nearness" of place a factor that might be considered is accessibility and therefore the types of information required to measure it. Accessibility is a function of available methods of transport and the properties of the transport routes. In general it appears to be most relevant when planning to visit the place or its neighbours. Clearly this may be relevant to some types of query and not to others. Support for accurate measurement of accessibility would require a network data structure with relevant impedance or cost factors attached to all links. If accessibility is a weak requirement then it may be that the approximate Euclidean distances derivable from limited coordinate data may serve as a surrogate measure.

2.4 Non-spatial concepts of place

If we assume that whenever a person specifies a place name they deposit some conceptual baggage that they associate with the name, then in modelling place it is reasonable to suppose that we should record attributes that may reflect the baggage. For example, if someone uses a named mountain range as a locator, then in expanding the search it is possible that they might regard hilly places bordering the mountains as more relevant than equally close neighbouring cities. If a named city were used as locator, it might be that neighbouring settlements that were in the same country as the named city were more relevant than those in another country that were equally near in Euclidean space.

In these examples, potentially relevant non-spatial attributes are topographic land-cover categories, and administrative (geopolitical) regions. If these types of attribute form the basis of regional hierarchies to which a place was referenced (by being inside or over-lapping an individual region) or of which it was a member, then the use of these hierar-chies for query expansion would automatically result in the inheritance of their properties by the places that are related to them. In the absence of an association with a regional hi-erarchy then it would appear important to attach classification terms to places in addition to the regional hierarchy relationships of containment, overlap and adjacency, and to be able to exploit them in search procedures.

Several authors have referred to the characteristics of place that offer opportunities and indeed constraints on the activities that may be performed there (Jordan et al., 1998). It is possible to envisage storing classification terms for place that reflect these opportu-nities or *affordances* directly. Alternatively and more economically it may be that certain types (such as port, mountain, river) may imply opportunities and actions and that for purposes of information retrieval the use of the classification terms to measure similarity with respect to class may serve as a rough surrogate for affordance.

3 MODELLING CONCEPTUAL TERMINOLOGY WITH THESAURI

There is long history of the use of thesauri in modelling terminology to assist in indexing and retrieval of information within particular domains. They are relevant to geographical information retrieval in that it may be possible to associate a specialised model of place, based on ideas referred to above, with existing thesauri representing non-spatial concepts associated with geographical locations. In this section we review briefly the principle types of relationships encoded within thesauri, in order to provide some background to our exploitation of thesaural relationships for purposes of non-spatial concept matching which is employed in association with place matching.

A major part of a thesaurus is usually one or more domain-specific classifications, derived either from a single source or resulting from a process of merging multiple clas-sifications within the same domain. Individual classification hierarchies may be grouped together within facets. One of the earliest suggested sets of facets is that of Ranganathan who proposed the five-fold division into Personality, Matter, Energy, Space and Time (PMEST) in the context of the Colon Classification system (Chan, 1994). The Art and Architecture Thesaurus (AAT) includes facets for physical attributes, styles and periods, agents, activities, materials and objects. The objects facet for example includes a Settle-ments and Landscapes Hierarchy which includes a variety of terms that express different types of place.

Classification structures are encoded in thesauri by means of generalisation and spe-cialisation relations, referred to as broader term (BT) and narrower term (NT). In order to denote their application to hierarchical encoding of generic relationships they may be described more specifically as BTG (Broader Term Generic) and NTG (Narrower Term Generic). Hierarchical relationships are also typically encoded in thesauri to describe part-whole relationships. Again broader and narrower relations are distinguished, in this case as BTP (Broader Term Partitive) and NTP (Narrower Term Partitive). Aitchison and Gilchrist (1987) distinguish four categories of part-whole relationships. These are: systems and organs of the body (e.g. ear BTP internal ear); geographical locations (USA BTP California); disciplines or fields of study (e.g. archaeology BTP marine archaeology); and hierarchical social structures (e.g. methodist church organisation BTP methodist district).

A third type of BT relationship is that of the instance relationship between an object and its class. The fourth type is that of polyhierarchical relationships, whereby some terms may be related to more than one parent class.

In acknowledging that users may refer to a very similar concept by means of different words, one term for a concept is usually designated the preferred term for purposes of encoding the term relationships. This leads to the converse equivalence relations of preferred-term and non-preferred-term that are referred to as the UF (or USE_FOR) and USE relationships. For example, if of the two terms 'lake' and 'mere', the former was the preferred, then the following relationships could be encoded in a thesaurus: *lake UF mere*; *mere USE lake*. This type of relationship is applicable to place name terminology as well as other concepts. Thus it allows multiple names for the same place to be referred to a single standard name. In the event of generating a unique name for a place that shares its name with other places, the unique name can be associated with the standard name via a USE relationship.

Clearly many terms are related to each other by relations other than the main hierarchical ones, while not being synonyms. This has given rise to the use of the associative relationships referred to as the related-term (RT) relationship. An important function of RT relationships is to link terms that may occur in separate facets but may be logically associated. For example the association *masons* RT *bricklaying* links terms in the Agents and Activities facets respectively of the AAT.

4 PLACE AND CONCEPTS IN OASIS

Some of the characteristics of place and concept described here have been implemented in OASIS (Ontologically Augmented Spatial Information System) to provide a basis for experimenting with geographical retrieval techniques. OASIS has been built using the Semantic Index System (Doerr and Fundulaki, 1998) which is an object-oriented hypermedia system that supports a number of semantic modelling constructs, a graphical user interface and API functions for data access. Schema creation can be performed with the TELOS language (Mylopoulos et al., 1990) or via data entry forms. Information can be classified at several levels including Token, Simple Class, and four levels of Meta Class. Both classes and objects are treated as objects and can have names, attributes and relationships to other objects.

The application area for which OASIS has been developed is that of cultural heritage and there is support therefore for the maintenance of archaeological artifacts that are linked to place by relationships of *found_at* and *made_at*. A Place type has been defined in OASIS as a subtype of Geographical Concept and its properties are illustrated in Figure 1. Places can be classified with one or more current and historical place types that, in our implementation, are mostly derived from the Art and Architecture Thesaurus (AAT). Of particular importance are place types that belong to regional hierarchies that can be expanded for purposes of information retrieval. The name of a place is designated via Standard Name and Alternative Name relations that include the attributes of *variant spelling*, *date* and *language*. Because of the possibility of duplication of names, places are given unique names that are referenced to their conventional name via the Standard Name relationship.

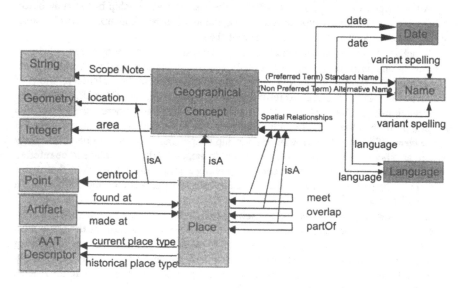

Figure 1: The schema for place in OASIS

The location of a place is represented quantitatively by a centroid (single point) defined by latitude and longitude values. It is treated as a specialisation of the location. The spatial representation of place includes the topological relations of *meets*, *overlaps* and *part-of* that are treated as specialisations of *spatial relationship*. These relationships are intended to allow an isolated place to be linked, by *part-of* or *overlap*, to a regional hierarchy, as well as serving to encode the structure of the hierarchies in terms of *part-of* and *meets* relations.

5   SEMANTIC DISTANCE MEASURES

5.1   Vector space methods

There are various methods for measuring the similarity between terms when matching a query expression with a target object. With vector space methods indexed documents are allocated coordinates in a multidimensional (term) space determined by the occurrence of terms in the document (Salton, 1989). A distance is then measured in vector space between a candidate document and the query expression, that is also located in the vector space. The approach is commonly applied to documents that may contain many terms, where the frequency of occurrence of individual terms may be taken into account in determining the location in vector space. The approach suffers from the disadvantage that query and target will only ever be regarded as similar if they share terms (at least) that are identical (either in full or when stemmed). It takes no account of the semantic relationships between terms that may be similar or related in meaning. It is intended for use with free text documents, rather than measurement of similarity between concepts that may be defined by a set of attributes and relationships.

5.2 Feature-based methods

In the feature matching methods introduced by Tversky (1977) an object is associated with a set of features which are compared with regard to their commonality and their difference. Similarity between two objects is measured as some function of the intersection of the features they have in common, the features unique to one of the objects and the features unique to the other object. By attaching different weights respectively to the features that belong to one object but not the other, a matching function can express the asymmetry commonly observed in relationships between objects where one is either more important or more prototypical than the other. Thus for example a house may be regarded as more similar to a building than a building is to a house, as a house is a subclass of building. Alternatively a settlement may be more similar to the state to which it belongs than is the parent state to the settlement.

5.3 Thesaural methods

Tversky's feature matching methods been shown to have potential for measuring similarity of spatial entities (Rodriguez et al., 1999), but they do have some limitations. One of these is that the results of matching operations will be skewed if one object has a different number of features than the other. The approach is intended for comparison of objects for which there are sets of descriptive features and consequently it cannot be applied directly to comparison of objects such as classification terms unless they are accompanied by a set of features. The approach also breaks down if there are differences in the terminology used to describe similar or equivalent features. Differences in terminology are widespread and may arise for example due to multiple organisations developing their own classification systems to refer to the same real-world domain. It is because of such differences in terminology that thesauri are widely used for indexing purposes. They provide a means for standardisation of terminology, for automatically identifying matches between equivalent terms via USE/UF relations, and for identifying terms that are similar, but not equivalent, in meaning by traversing the hierarchical and associative relations.

The use of thesauri or similar semantic nets has led to the development of various semantic distance measures based on the traversal of the semantic relationships. A simple approach is to base the semantic distance between two terms within a thesaurus on the shortest path between them (Rada et al., 1989; Lee et al., 1993). In a classification hierarchy this is the smallest number of is-a links between the two terms. A variation on the method is to attach different weights to links according to their type or their depth in the hierarchy (Kim and Kim, 1990; Richardson et al., 1996; Tudhope and Taylor, 1997). In OASIS this approach, with weighted links, has been applied to the determination of similarity between AAT terminology that we have used to define non-spatial concepts. We use the following formula to determine the thematic distance TD between two terms a and b:

$$\text{TD}(a, b) = \frac{C_{a,x_1}}{L_{x_1}} + \frac{C_{x_1,x_2}}{L_{x_2}} + \frac{C_{x_2,x_n}}{L_{x_n}} + \dots + \frac{C_{x_n,b}}{L_b} \tag{1}$$

It is based on a summation of the weighted links in the shortest path from a to b. $C_{j,k}$ is the weight of the relationship between intermediate terms j and k in the shortest path between a and b, and is related to the thesaural type of the relationship. L_i is, by default, the hierarchical level of the term i, hence resulting in smaller distances between terms lower down a hierarchy. An example of the application of the method is given in section 5.7.

5.4 Non-common super-classes

An alternative approach to measuring similarity of classification terms, in the context of a thesaurus or some other semantic net that includes hierarchical relationships, is one based on the non-common super-classes of pairs of terms (Spanoudakis and Constantopoulos, 1994). The non-common super-classes of two objects a and b consist of parent classification terms that belong to a but not to b and those that belong to b but not to a. These terms may be regarded as analogous to the distinctive features of Tversky's methods. While the feature-based methods include an explicit measure of the common features, this is implicit in the non-common super-classes method, since the semantic net encodes relations of class generalisation or of part-whole directly, so that by definition if a pair of terms has no non-common super-classes then they must be closely related within the semantic space of the ontology. A further difference from the feature-based methods is the use of level-specific values whereby differences between terms decrease with increasing depth in the hierarchy, just as in the shortest distance methods referred to above.

5.5 An hierarchical spatial distance measure

In our treatment of place we regard the non-common super-classes method as applicable to the measurement of similarity between places with regard to the regional hierarchies to which they belong, via *part-of* or *overlap* relations. It is considered appropriate as it leads to measures of similarity that reflect differences in inherited properties of place as determined by the multiple hierarchies to which a particular place may belong. A limitation of the non-common super-classes method compared to the feature-based method is that it cannot express asymmetry of similarity. In order to address this shortcoming we propose adapting the method by including separate weights α, β for the distinctive super-classes of the two terms respectively. We also introduce a further weighting term γ to provide flexibility with regard to inclusion of the query and candidate terms in the measurement formula. The hierarchical distance HD of query place a from candidate place b is

$$\text{HD}(a, b) = \sum_{x \in \{a.PartOf - b.PartOf\}} \frac{\alpha}{L_x} + \sum_{y \in \{b.PartOf - a.PartOf\}} \frac{\beta}{L_y} + \sum_{z \in \{a,b\}} \frac{\gamma}{L_z} \tag{2}$$

where L_x and L_y are the hierarchical levels of the distinctive super-parts of a and b respectively, while L_z are the hierarchical levels of a and b. $a.PartOf$ and $b.PartOf$ refer to all super-parts of a and b respectively, i.e. at all higher hierarchical levels. If a and b are to be included in the measurement then γ takes on a non-zero value, otherwise it is zero. If either a is a sub-part of b or b is a sub-part of a (separated by one or more hierarchical levels) then α is set larger than β. Otherwise α and β are equal. Thus in a measurement of the distance between a and b, if a is the super-part it will have no non-common super-parts and hence the distance will be biased by the smaller weight. Conversely if a is a sub-part of the candidate term, its non-common parents (that include b) will be biased by the larger weight, resulting in a greater distance value.

It is envisaged that γ should be set non-zero when both a and b are members of the regional hierarchies, as opposed to one of them simply being referenced to a member of a hierarchy. Thus two sub-regions with a common parent will be separated by a finite distance, reflecting the fact that they are not the same region. However if two non-regional places belong to the same parent region then, purely with regard to the regional hierarchy, there is no difference between them. Clearly they will have a difference in Euclidean space and they may have a difference with regard to their individual place classes.

5.6 Coordinate-based distance

As indicated in the description of the OASIS schema for place, we attach a centroid to each place consisting of two coordinates. In order to support global applications, we encode the coordinates as latitude and longitude. Earth surface distances are then calculated along great circles. We refer to this measure as Euclidean Distance. In using centroids, the resulting distances may be regarded as somewhat error prone, particularly in the case of places with considerable area extent. As is explained in Alani et al. (2001), centroids can be used to approximate the boundaries of regions provided there is knowledge of both contained and neighbouring external places that are associated with centroids. The approximated boundaries may then be used to determine distance between boundaries or between points and boundaries.

5.7 Examples

In this section we illustrate the use of thematic distance and the hierarchical distance measure and show how they can be combined with Euclidean distance to produce an integrated ranking measure.

5.7.1 Thematic distance

As explained in the previous section, we have applied the thematic distance (TD) measure to comparison of concepts that belong to the AAT. The costs C of traversal of the different types of relationships have been set to BT 3, NT 3 and RT 4. When traversing RT relationships the level is taken as that of the originating term rather than the destination term of a relationship.

Referring to the example in Figure 2, there are two possible paths between axes (weapons), at level 4, and throwing axes, at level 6. Thus:

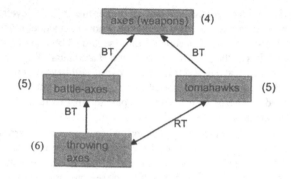

Figure 2: Thematic distance example using terms from the AAT

$$TD(axes\ (weapons), throwing\ axes) = \frac{C_{NT}}{level\ of\ battle\ axes} + \frac{C_{NT}}{level\ of\ throwing\ axes}$$

$$= \frac{3}{5} + \frac{3}{6} = 1.1$$

Note that NT relationships are simply the converse of BT relationships. The second path produces the distance:

$$TD(axes\ (weapons), throwing\ axes) = \frac{C_{NT}}{level\ of\ tomahawks} + \frac{C_{RT}}{level\ of\ throwing\ axes}$$

$$= \frac{3}{5} + \frac{4}{5} = 1.4$$

Since it is the first path that has the lowest cost the value of TD in this case is 1.1. For a more detailed treatment of RT relationships see Tudhope et al. (2001).

5.7.2   *Hierarchical distance measure*

We illustrate the use of the hierarchical distance measure with regard to an example scenario in Figure 3, in which several places of type hill are associated with members of an administrative regional hierarchy. In order to illustrate the application of a polyhierarchy, the association between hills and administrative regions represents both part-of and overlap relationships. The regional hierarchy is built entirely from part-of relationships.

In this scenario Scotland is placed at hierarchical level 4 (Scotland *part of* United Kingdom *part of* Europe *part of* World), its sub-regions are at level 5, and the hills are therefore at level 6. In the following examples of distances between hills the weights α and β have been set equal to 1, while γ has been set to zero, giving the following results:

1. $HD(Henshaw\ Hill, West\ Cairn\ Hill) = 0$

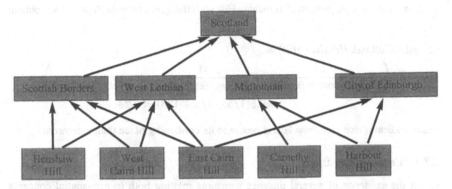

Figure 3: Example of hill places referenced by part-of and overlap relationships to an administrative hierarchy

reflecting the fact that the two places both overlap the same two regions of Scottish Borders and West Lothian and no other regions.

2. $\text{HD}(\textit{Henshaw Hill}, \textit{East Cairn Hill}) = \dfrac{1}{\textit{level of City of Edinburgh}}$

$$= 1/5 = 0.2$$

reflecting the fact that East Cairn Hill overlaps the City of Edinburgh, but Henshaw Hill does not.

3. $\text{HD}(\textit{Henshaw Hill}, \textit{Carnethy Hill}) =$

$$\frac{1}{\textit{level of Scottish Borders}} + \frac{1}{\textit{level of West Lothian}} + \frac{1}{\textit{level of Midlothian}}$$

$$= (1/5 + 1/5) + 1/5 = 0.6$$

4. $\text{HD}(\textit{Henshaw Hill}, \textit{Harbour Hill}) = (1/5 + 1/5) + (1/5 + 1/5) = 0.8$

To illustrate the application of asymmetry, the values of α and β may be set to 1 and 0.5 respectively. When Scotland is compared with the query term Henshaw Hill we obtain the following result:

5. $\text{HD}(\textit{Henshaw Hill}, \textit{Scotland}) =$

$$1\left(\frac{1}{\textit{level of Scottish Borders}} + \frac{1}{\textit{level of West Lothian}} + \frac{1}{\textit{level of Scotland}}\right) + 0$$

$$= (1/5 + 1/5 + 1/6) = 0.57$$

whereas with the comparison of Henshaw Hill with the query term of Scotland we obtain:

6. $HD(Scotland, Henshaw\ Hill) =$

$$0 + 0.5 \left(\frac{1}{level\ of\ Scottish\ Borders} + \frac{1}{level\ of\ West\ Lothian} + \frac{1}{level\ of\ Scotland} \right)$$
$$= 0.5(1/5 + 1/5 + 1/6) = 0.28$$

which indicates that Henshaw Hill is nearer to its containing place than *vice versa*.

5.7.3   Combining measures

Given the existence of several distance measures relating both to non-spatial concepts and to qualitative and quantitative ("Euclidean") space, some means is required to combine such measures to provide a single overall ranking. In experiments with OASIS, the Euclidean and hierarchical distance measures were combined by normalising the individual measures before applying weights to the two components to produce a total spatial distance (TSD) measure defined as

$$TSD = (w_e ED_n + w_h HD_n) \qquad (3)$$

where ED_n and HD_n are the normalised measures and w_e and w_h are the respective weights that sum to one. The TSD may then be combined with a normalised thematic distance as follows:

$$Score = 100 - (w_t TD_n + w_s TSD_n) \qquad (4)$$

to produce a value between 0 and 100 where w_t and w_s are the weights for theme and space that also sum to one. Table 1 illustrates an example of applying the score to rank the results of a query for "axes in Edinburgh," where axes has been specified as belonging to the weapons hierarchy. In this case the weights were set as follows:

$$Score = 100 - (0.4 * TD_n + 0.6 * (0.6 * ED_n + 0.4 * HDn)) * 100$$

In the experiment, asymmetry was not taken into account in the hierarchical distance measure. Due to a paucity of real data in some geographic regions, some imaginary data items were added, referring for example to tomahawks. Taking the example of an occurrence of tomahawks (weapons) in the region of Currie (a part of Edinburgh in the administrative regional hierarchy), the normalised thematic distance between tomahawks and axes (weapons) was 0.428, while the normalised Euclidean distance and hierarchical distances values of Currie from Edinburgh were 0.321 and 0.615 respectively. This results in a calculation of the score of:

$$Score = 100 - (0.4 * 0.428 + 0.6 * (0.6 * 0.321 + 0.4 * 0.615)) * 100 = 57\%$$

In the example it is apparent that places of Edinburgh and the contained places of Edinburgh are ranked before neighbouring places outside Edinburgh such as East Lothian.

The ranking has resulted in some cases in the regional hierarchy modifying ranking that would be produced with Euclidean distance alone. Note also for example that because throwing axes are semantically more distant from axes (weapons) than are tomahawks, occurrences of the former that were found in Edinburgh are relegated to a lower score than tomahawks.

ID	ARTEFACT	PLACE FOUND	SCORE
AF303	Axes (weapons)	Edinburgh'Edinburgh	100
AF399	Axes (weapons)	Edinburgh'Edinburgh	100
DE121	Axes (weapons)	Edinburgh'Edinburgh	100
AT339	Tomahawks (weapons)	Edinburgh'Edinburgh	83
AT333	Tomahawks (weapons)	Edinburgh'Edinburgh	83
AT340	Tomahawks (weapons)	Edinburgh'Edinburgh	83
AF340	Axes (weapons)	Edinburgh'Leith	81
AF331	Axes (weapons)	Edinburgh'Leith	81
AF432	Axes (weapons)	Edinburgh'Corstophine	79
AF434	Axes (weapons)	Edinburgh'Luddingston	78
AF334	Axes (weapons)	Edinburgh'Currie	74
AF332	Axes (weapons)	Edinburgh'Currie	74
AF341	Axes (weapons)	Edinburgh'Currie	74
AF321	Axes (weapons)	Edinburgh'Dalmeny	70
AF329	Axes (weapons)	Edinburgh'Ratho	69
AF349	Axes (weapons)	Edinburgh'Ratho	69
AF335	Axes (weapons)	Edinburgh'Kirkliston	68
AF339	Axes (weapons)	Edinburgh'Kirkliston	68
AF337	Axes (weapons)	Edinburgh'Kirkliston	68
TA361	Throwing axes	Edinburgh'Edinburgh	60
TA362	Throwing axes	Edinburgh'Edinburgh	60
AF510	Axes (weapons)	East Lothian'Musselburgh	60
AF429	Axes (weapons)	East Lothian'Inveresk	59
AF449	Axes (weapons)	East Lothian'Inveresk	59
AT390	Tomahawks (weapons)	Edinburgh'Currie	57
AF499	Axes (weapons)	Midlothian'Dalkeith	56
AF456	Axes (weapons)	Midlothian'Borthwick	56
AF229	Axes (weapons)	West Lothian'Kirknewton	54

Table 1: Example of ranking the results of a query for "axes (weapons) in Edinburgh"

6 CONCLUDING REMARKS

This paper has addressed the problem of developing facilities for geographical information retrieval in which the user may employ place names and concept terms that may not match precisely with the terms used to describe information of interest. A model of place has been proposed in combination with semantic closeness measures that can be used to rank the relevance of retrieved information with regard to the user's query terms. The model of place adopts a parsimonious approach to storage of spatial data with a view to providing potentially global coverage of geographic place names. The place names are associated with alternative versions of their name and one or more place type categories. Instances of place are linked to other places via qualitative spatial relations and are linked to geographical coordinate space with a single centroid. A hierarchical spatial distance measure is introduced that determines the distance between two places in terms of the number of non-common, parent places to which they belong, as determined by relationships of containment and overlap. The measure is combined with Euclidean distance to create an integrated spatial distance measure. A semantic distance measure based on weighted shortest paths within a thesaurus of classification terms is combined with the spatial measures to obtain an overall ranking of the results of queries that specify a thematic query term in combination with a place name. The techniques presented are intended to make some progress towards handling natural language terms in geographical queries. Some preliminary user experiments, not reported here, have given support to the validity of the methods presented. This paper has focused on the use of place name as a locator and the inherent thematic or non-spatial aspects of place have only played a significant role with regard to measurement of semantic distance on the basis of non-common parent places. There is clearly scope to employ an explicit thematic distance measure that uses the place type terms to provide more sensitive distinctions between place, with regard to cultural, socio-economic and historic perspectives, than that provided by the parent places. This would be of particular importance if the purpose of a query were to find places similar to a specified query place, as opposed to finding some phenomena that are located at a specified place. There is also scope for experimenting with a wider variety of spatial closeness measures. It would be possible, for example, to employ qualitative measures of distance based on contiguity of neighbouring places (Jones et al., 1996) and it would also be possible to weight distance measures according to degrees of overlap between places following the approach of Beard and Sharma (1997).

The methods presented here have been motivated by problems of information retrieval, but they have a wider application. In particular the semantic closeness measures may have potential for assisting in solving problems of geographical data integration in which data from different sources may employ different classification terminology and different place names to refer to similar locations. In these contexts the similarity measures could be used to help identify the equivalence of multiple representations of the same real-world phenomena.

ACKNOWLEDGMENTS

We would like to thank the J. Paul Getty Trust and Patricia Harpring in particular for provision of their TGN and AAT vocabularies; Diana Murray and the Royal Commission on the Ancient and Historical Monuments of Scotland for provision of their dataset; and Martin Doerr and Christos Georgis from the FORTH Institute of Computer Science for assistance with the SIS.

REFERENCES

Aitchison, J. and Gilchrist, A. (1987). *Thesaurus Construction: A practical manual.* London: Aslib.

Alani, H., Jones, C. B., and Tudhope, D. S. (2001). Voronoi-based region approximation for geographical information retrieval with gazetteers. *International Journal of Geographical Information Science*, 15(4):287–306.

Aurenhammer, F. (1991). Voronoi diagrams—a survey of a fundamental geometric data structure. *ACM Computing Surveys*, 23(3):345–405.

Beard, K. and Sharma, V. (1997). Multidimensional ranking for data in digital spatial libraries. *International Journal of Digital Libraries*, 1:153–160.

Chan, L. (1994). *Cataloging and Classification: An Introduction.* New York: McGraw-Hill, 2nd edition.

Couclelis, H. (1992). Location, place, region and space. In Abler, R. F., Marcus, M. G., and Olson, J. M. (Eds), *Geography's Inner Worlds*, pp. 215–233. New Jersey: Rutgers University Press.

Curry, M. R. (1996). *The Work in the World—Geographical Practice and the Written Word.* Minneapolis: University of Minnesota Press.

Doerr, M. and Fundulaki, I. (1998). SIS-TMS: A thesaurus management system for distributed digital collections. In Nikolaou, C. and Stephanidis, C. (Eds), *Second European Conference on Research and Advanced Technology for Digital Libraries, ECDL'98*, pp. 215–234. Crete, Greece: Heraklion.

Gould, P. and White, R. (1986). *Mental Maps.* London: Allen and Unwin.

Guarino, N., Masolo, C., and Vetere, G. (1999). OntoSeek: Content-based access to the web. *IEEE Intelligent Systems*, 14(3):70–80.

Harpring, P. (1997). Proper words in proper places: The thesaurus of geographic names. *MDA Information*, 2(3):5–12.

Hill, L. L., Frew, J., and Zheng, Q. (1999). Geographic names: Implementation of a gazetteer in a georeferenced digital library. *D-Lib Magazine*, 5(1). http://www.dlib.org/dlib/january99/hill/01hill.html.

Johnson, R. J. (1991). *A Question of Place: Exploring the Practice of Human Geography.* Blackwell.

Jones, C. B., Taylor, C., Tudhope, D., and Beynon-Davies, P. (1996). Conceptual, spatial and temporal referencing of multimedia objects. In Kraak, M. J. and Molenaar, M. (Eds), *Advances in GIS Research II*, pp. 33–46. London: Taylor & Francis.

Jordan, T., Raubal, M., Gartrell, B., and Egenhofer, M. J. (1998). An affordance-based model of place in GIS. In Poiker, T. K. and Chrisman, N. (Eds), *Proceedings 8th International Symposium on Spatial Data Handling*, pp. 98–109. International Geographical Union.

Kim, Y. W. and Kim, J. H. (1990). A model of knowledge based information retrieval with hierarchical concept graph. *Journal of Documentation*, 46(2):113–136.

Larson, R. R. (1996). Geographic information retrieval and spatial browsing. In Smith, L. and Gluck, M. (Eds), *Geographic Information Systems and Libraries: Patrons, Maps, and Spatial Information*, pp. 81–124. Urbana-Champaign: University of Illinois, GSLIS.

Lee, J. H., Kim, M. H., and Lee, Y. J. (1993). Information retrieval based on conceptual distance in IS-A hierarchies. *Journal of Documentation*, 49(2):113–136.

McCurley, K. (2001). Geospatial mapping and navigation on the web. In *Proceedings WWW10*. http://www10.org/cdrom/papers/278.

Moss, A., Jung, E., and Petch, J. (1998). The construction of WWW-based gazetteers using thesaurus techniques. In Poiker, T. K. and Chrisman, N. (Eds), *Proceedings 8th International Symposium on Spatial Data Handling*, pp. 65–75. International Geographical Union.

Mylopoulos, J., Borgida, A., Jarke, M., and Koubarakis, M. (1990). Telos: A language for representing knowledge about information systems. *ACM Transactions on Information Systems*, 8(4):325–362.

Rada, R., Mili, H., Bicknell, E., and Blettner, M. (1989). Development and application of a metric on semantic nets. *IEEE Transactions on Systems, Man and Cybernetics*, 19(1):17–30.

Relph, E. (1977). *Place and Placelessness*. Pion Limited.

Richardson, R., Smeaton, A. F., and Murphy, J. (1996). Using WordNet for conceptual distance measurement. In Leon, R. (Ed), *Information Retrieval: New Systems and Current Research*, volume 2, pp. 100–123. Taylor Graham.

Rodriguez, M. A., Egenhofer, M. J., and Rugg, R. D. (1999). Assessing semantic similarities among geospatial feature class definitions. In Vckovski, A., Brassel, K., and Schek, H. J. (Eds), *Interop'99*, volume 1580 of *Lecture Notes in Computer Science*, pp. 189–202. Berlin: Springer-Verlag.

Salton, G. (1989). *Automatic Text Processing: The Transformation, Analysis, and Retrieval of Information by Computer*. Reading, MA: Addison-Wesley.

Tuan, Y. (1977). *Space and Place: The Perspective of Experience*. Edward Arnold.

Tudhope, D., Alani, H., and Jones, C. (2001). Augmenting thesaurus relationships: Possibilities for retrieval. *Journal of Digital Information*, 1(8). http://jodi.ecs.soton.ac.uk/Articles/v01/i08/Tudhope/.

Tudhope, D. and Taylor, C. (1997). Navigation via similarity: Automatic linking based on semantic closeness. *Information Processing and Management*, 33(2):233–242.

Tversky, A. (1977). Features of similarity. *Psychological Review*, 84(4):327–352.

Walker, D., Newman, I., Medyckyj-Scott, D., and Ruggles, C. (1992). A system for identifying datasets for GIS users. *International Journal of Geographical Information Systems*, 6(6):511–527.

CHAPTER 12

Placing Cultural Events and Documents in Space and Time

Ray R. Larson

School of Information Management and Systems
University of California, Berkeley, CA, 94720, USA

1 INTRODUCTION

Digital library projects throughout the world have been creating very large-scale repositories of digital information on a wide range of topics. The contents of digital libraries today include collections of scientific data, digitized versions of classical and modern texts in the original languages, commentaries on those texts, historical and current social science survey information, collections of digitized video and audio, and collections of images—including digitized paintings, sculpture, drawings, and other museum objects. In short, the cultural heritage of the world is rapidly being replicated in digital form.

Digital library research has become concerned with this steadily increasing range of genres and materials and, more challengingly, with the use of diverse digital genres in conjunction with each other. For example, researchers associated with the Electronic Cultural Atlas Initiative (ECAI, 2001) are investigating means of combining textual information with geospatial data, enabling cultural, historical, and social data to be represented in time and space through Geographical Information Systems (GIS). Linking the mention of place names to other geo-spatial information involves three different genres: toponym-rich texts, GIS databases, and, mediating between the two, gazetteers, made up of structured records about toponyms (place names) and the geographic coordinates of the places referred to by those names.

The characteristics of digital information permit repositories of geo-referenced digital information be indexed and searched in ways that were never possible in a print-on-paper environment. Users of digital libraries, whether scholars or ordinary citizens, often will have needs for information that are best approached from a geo-temporal perspective. These users include historians who require information on specific areas (at particular times), social scientists tracing trends over time within a particular society or country, students studying particular places and events as well as citizens with interests in current and historic places and times.

This paper will examine the notion of *geographic information retrieval* in the context of digital libraries focusing on the needs and purposes of cultural and humanities applications. We will examine the issues of indexing and retrieval for materials with geographic content or associations. We will also examine some of the characteristics of geo-referenced information and how such information is incorporated into digital libraries, and used by scholars in their work. The intent is to show how such geo-referenced information is being used to enrich scholarly work and to provide public information about historic and cultural events. In addition we will examine some issues in the design and use of digital libraries with regard to retrieval and access to geographically oriented contents. The next section describes the primary components for geographic access and retrieval within a digital library and describes the characteristics of geographic information retrieval and spatial querying in a digital library context. Following sections examine indexing and access creation for geo-referenced sources, both manual and automated, and how such information is being created and used in various scholarly endeavors (focusing on current work of the Electronic Cultural Atlas Initiative). Finally, the conclusion will examine some general issues and characteristics of geo-referenced multimedia information systems, and how they might evolve in the future.

2   GEOGRAPHIC INFORMATION RETRIEVAL: COMPONENTS AND METHODS

Geographic information retrieval (GIR) is an applied research area that combines aspects of DBMS research, user interface research, GIS research, and information retrieval (IR) research, and is concerned with indexing, searching, retrieving and browsing of geo-referenced information sources, and the design of systems to accomplish these tasks effectively and efficiently (Larson, 1996). It is intended to be an inclusive term that implies all of the concerns of traditional IR research with an emphasis, or addition, of spatially and geographically oriented indexing and retrieval.

Information retrieval research has always taken a probabilistic approach to providing access to documents. By document we mean any item of potential interest to a user in a collection or database, regardless of the content or form (e.g., text, images, maps, video). IR is concerned with matching a user's need for information to the items in the database most likely to satisfy that need (that is, to provide, to the greatest extent possible, all and only the relevant documents). The sort of exact matching typical of data retrieval operations in DBMS is not sufficient for IR, at best there is a largely subjective and indeterminate matter of whether or not a document satisfies a user's need for information. In GIR we must be concerned with both deterministic retrieval (such as finding all data sets containing coordinates within an area of interest) and probabilistic or "fuzzy" retrieval (such as finding all towns "near" a major river).

Four primary and necessary components may be found in any GIR system, such as a digital library with geo-referenced information. These components are:

a. *Database*: The data itself, including all of the information sources whether or not they include explicit geo-references.

b. *Indexes* to the searchable metadata elements of the database. Not all elements of a database may be searchable in some systems, but only preselected elements considered by the designers of the database to be the appropriate or commonly used elements. Indexing is required for both efficient access to large databases, and to organize and limit the set of elements of a database that are accessible. Most information retrieval systems derive their index elements from the contents of the items to be indexed. The derivation may be simple extraction (such as extracting keywords from a text), inferential extraction (such as mapping from text word to thesaurus terms) or it may be intellectual analysis and assignment index items (such as assigning subject headings to a document). In GIR there may be direct extraction of coordinate or coverage information from data sets, as well as intellectual indexing (such as assignment of bounding box coordinates to an aerial photograph), and inferential indexing (assignment of coordinates for places mentioned in a text).

c. *Retrieval and matching algorithms*: In retrieving items from the database the system will use algorithms to match the query to the index elements (and hence to the database contents). These algorithms typically will belong to a class of retrieval algorithms that are probabilistic in nature, and may involve the actual calculation of probabilities and use of statistical inference methods. Or they might take another approach based on differing models of the document space (such as Salton's, 1989, vector space model). The principle goal of these algorithms, regardless of the model used, is to attempt to find all of the (partial) matches between query and document and to rank them based on some measure of "goodness," so that the "best" matches receive the highest ranks. Database retrieval methods, on the other hand, are deterministic and therefore demand an exact match between the query specification and the contents of the database. In GIR both approximate, partial matching and strict deterministic matching are of value in processing geographic and spatial queries.

d. *User interface*: The user interface provides the user with a way to specify a query, to review the results of the query, and (possibly) to modify and re-direct the query based on previous results. In GIR the interface will usually include a map representation of the area under consideration (either as a primary method of retrieval, or as a way of restricting the query results to those that within the area displayed). In some systems there may also be similar representations for temporal specification or qualification (such as a "time-line" with a sliders for beginning and end of the period of interest).

In GIR systems, each of these components require features and operations not found in traditional text-based IR systems. Most of the features have to do with support for the specification and efficient searching of geographically based information. For example, the index component may need to use access methods optimized for spatial retrieval (such as R-Trees) in place of the B-Tree and inverted file methods typically employed in text retrieval systems in order to provide efficient access to spatial information. Frank (1991) has discussed the characteristics of geographic data that require special access methods and data structures. We will use the terms "geographic queries" and "spatial queries" somewhat interchangeably in this discussion. Both terms imply querying a spatially indexed database based on relationships between particular items in that database within a particular coordinate system (or compatible coordinate systems). Spatial querying is the more general term. It can be defined as queries about the spatial relationships (intersection, containment, boundary, adjacency, proximity) of entities geometrically defined and located in space (De Floriani et al., 1993) without regard to the nature of the coordinate system. Geographic querying assumes that the space is delineated by the well-defined coordinate systems of the "real world." We will concentrate on geographic querying, although the underlying implementation might be a spatial database system rather than a geographic information system. We will not examine access methods here, but will concentrate on a basic classification of types of spatial queries.

Spatial relationships may be both geometric and topological (spatially related but without measurable distance or absolute direction). Examples of topological relations include such properties as adjacency, connectivity, and containment. For example, whether some place is inside or outside of the county of Essex has to do with a relationship to an arbitrary boundary, but the distance or direction between the boundary and the location is not important. In historical GIR applications this type of relationship becomes more complex because of the changes in boundary definitions over time. A building within today's county of Essex may have been in a different county 500 years ago. Spatial and geographic queries can combine both geometric and topological elements. Frew et al. (1995) suggest that there are two primary classes of requests from users, the "What's here" query and the "Where's this" query. The first type of query stems from a desire to discover what information is available about a particular location, while the second stems from a desire to find out where certain phenomena occur. Within this simple classification of spatial and geographic queries, there are a number of different types of queries, distinguished by how the locations of interest are defined. The classification is based on the types of spatial queries defined by Laurini and Thompson (1992) and De Floriani et al. (1993) (adapted from the discussion in Larson, 1996).

The types of spatial queries submitted by users to a GIR system may be arbitrarily complex in the types of information desired, the limitations on the areas, time periods, etc. covered, and many other conditions (spatial or not) that might be specified in such a query. If we concentrate on only the spatial or geographic aspects of the query, there are a number of query types that can be distinguished based on the types of information provided by the user in the query. These are:

Point-in-polygon queries: The Point-in-polygon query asks the question "What information is available for this X,Y point in the current coordinate system?" In GIR the implication is that the specified point is contained within the polygonal boundaries or "footprint" of some set of geo-referenced datasets. The proper response to such a query is to provide a listing or description of the datasets containing the specified point. The query essentially asks for any geo-referenced object or geographic dataset that contains, surrounds or refers to a particular spot on the surface of the earth.

Region queries: A Region query asks "What information is available for this user-defined region?" Instead of referring to a particular point in the coordinate space, a region query defines a polygon in that space and asks for information regarding anything that is contained in, adjacent to, or overlaps the polygonal area so defined. There are a number of potential variants or restrictions that might be applied, for example, a user might be asking "Which point encoded items lie within the region," "What lines (borders, rivers, etc.) lie within or the cross the region," "Which areas (or regional datasets) overlap this region," "Which areas (or regional datasets) lie entirely within this region," or "Which areas share a border with this region." Any combination of elements or containment criteria might be specified, given the needs of the particular searcher. In addition, the specified query region can be any polygon, ranging from regular shapes such as rectangles or even circles (which would be the same as a Buffer Zone query on a point as discussed below), to irregular shapes like the boundary of a city, or any arbitrary set of points defining a closed polygonal shape. The containment criteria need not be precise, but may use "fuzzy" or probabilistic interpretations of such things as the maximum or minimum areas of overlap for an object to be considered included in the specified area, or the coverage areas for particular datasets that are candidates for retrieval (Brimicombe, 1993).

Distance and buffer zone queries: The distance and buffer zone query asks "what information is available within some fixed distance of this object (point, line or polygon)." For example, "What towns are situated within 5 kilometers of border of Wales and England?", "What mines are within 5 miles of this city?", "What information is there on the occupations of people living within 5 kilometers of the River Avon?", etc. The buffer zone specified need not be exact, e.g., "what datasets describe the area near this point?" and inclusion can be considered a fuzzy or probabilistic function based on the location of the database objects. For such queries, a ranked list of database objects ordered by "nearness" to the specified point, might be a better response than an arbitrary definition of a distance.

Path queries: Path queries are a somewhat more specialized form of spatial query that require the presence of a network structure in the spatial or geographic data. Networks are simply sets of interconnected line segments, representing such things as roads, oil or water pipelines, etc. A typical sort of path query involves finding the shortest route from one point in the network to another. For example, a path query might ask "What is the shortest train route from Liverpool to Edinburgh?" Note that path queries can become more complex (and uncertain) multimedia queries when criteria other than distance or direction are involved in the query.

"Multimedia" queries: Queries where a combination of multiple geo-referenced sources are used and combined to specify the acceptable results is considered a "multimedia query" by Laurini and Thompson (1992). Several of the examples used above would actually be considered multimedia queries, because they combine multiple geo-referenced information sources in resolving a query. For example, resolving the query about the occupations of people living near the Avon would require, at least, geo-referenced occupational data and the polygonal data for the river itself. Multimedia queries may include information without direct geo-referencing (including cadastral information, such as ownership records for particular parcels of land defined by street address and parcel size). Multimedia queries can become quite complex, and may require combining aerial or satellite photographs or remote sensing data, map and cadastre information, often from different databases, with different measurements, scales and levels of detail. Any multimedia information system may include a wide variety of spatial and non-spatial information that may have a geographic association, if not a precise location (see, for example, Griffiths, 1989).

The types of queries discussed above can be combined in a complex search. For example, "What streams and rivers flow through the county in which the city of Colchester (UK) exists" would require a point-in-polygon search (or possibly a gazetteer lookup) of county information to locate the county containing the city, and a region search to identify the streams and rivers that intersect the county area.

For GIR systems to support historical information as well as current data, the other obvious attribute to include in query specifications is some indication of the time period or coverage to be included in the results. Information on, for example, county boundaries will be quite different depending on the time period. For example, the Historical Geographic Information System for Great Britain (GBHGIS, 2001), maintains a record of boundary changes, extracted from a combination of old maps and also textual records describing changes, all maintained on continuous time dimension, with changes recorded by date. The creators of the GBHGIS describe how one official report shows that a particular district had its boundary changed on the 1st of January 1905, while two separate maps show boundaries in 1901 and 1908. The GBHGIS system stores all of this information, including the date of the change, so it can construct an accurate map of the district for any date, not just 1901 and 1908. The associated Great Britain Historical database (GB-HDB) provides a large integrated database of geographically-located historical statistics for Great Britain, mainly drawn from the period 1851–1939. The compilers have worked to integrate the spatial and temporal references in data ranging from boundary information and gazetteers to census data and marriage statistics to photographs and traveler's journals.

This is in contrast to the conventional applications of database systems where the temporal dimension is important. Many of these (including DBMS applications such as banking account processing, where the date and sequence of transactions is critical) deal only with simple discrete changes occurring at a single point in time. They are seldom, if ever, concerned with continuous transitions of, e.g. riverbeds, over time. As Peuquet (2001) points out "There are two basic types of spatial change relating to objects; they can (a) change location in space, i.e., move, and they can (b) change in spatial extent by growing or shrinking and by changing shape. Depending on the application, change can consist of movement of point objects, which have no notable areal extent. This might include applications such as the tracking of delivery vans, taxi cabs, or military vehicles." Many of the types of changes that need to be tracked and represented for historical work consist of type "b", and may also include items that combine both "a" and "b" type changes over time. Although such changes might be considered as continuous, most historical information, as the GBHGIS discussion above indicates, like cadastral information is typically discrete, involving a transaction at a particular point in time where, for example, a parcel of land changes hands from one owner to another, or boundaries of a region are surveyed and measured. Other, truly continuous changes, such as incremental deposition of land on a river delta, silting and dredging of harbors, etc. may also have an significant influence on historically interesting information such as habitation patterns, economic conditions of an area, and so on. However for the purposes of historical research, it will usually be the case that such information is seldom of an accuracy or detail (or temporal regularity) to warrant more true spatio-temporal modelling and representation such as those suggested by Hazarika and Cohn (2001).

To support the cultural and humanities-oriented work of scholars, GIR systems should combine the text or concept-based retrieval associated with conventional information and database systems with the kinds of spatial and temporal queries discussed above. Scholars need to be able to locate relevant information, where the relevance takes into account explicit or implicit spatial and temporal constraints. Providing today's county boundaries will give incorrect information on population, land use, tax roles and other information if the period of interest to the scholar is the 1750's.

One approach to combining relevance ranking with spatial ranking is discussed by Jones et al. (this volume). Their approach is to use a ranking of documents (database entries) based on the proximity of the places named in query to the places named in the document text (or database entry), while taking into account containment relations for the places named in the query. This is used in conjunction with a thesaurus-based conceptual proximity measure between the terms used in the query and the conceptually related terms used in the documents themselves. The spatial and non-spatial conceptual components of the ranking are combined using a weighted linear equation giving different weights to each component. At Berkeley we are working on similar methods of combining elements for information retrieval based on logistic regression methods using the Cheshire II system. In our case we commonly use XML or SGML documents with many place names occurring in any document. Part of the document indexing and ranking (still under development) is concerned with disambiguating and ranking documents or parts of documents based on combinations of spatial information and statistical information about term usage in the documents and collection (i.e., conventional IR methods).

Searching a geographically indexed database or digital library is an activity that assumes the searcher has a good notion of what he or she wants, and is able to specify that need in some form. Most of the queries used as examples in the above discussion reflect this. Another type of "searching" is much less directed, and while it assumes that the users have some notion of the type of information desired, they may not be able to specify that information in a query language. What is needed in such cases is (in effect) the ability to navigate the database geographically and temporally, without requiring explicit query formulation. This "spatial browsing" combines ad hoc spatial querying with interactive displays of digital maps to permit the user to explore the geographical dimension of information in a database or digital library. In such browsable displays, representations of the coverage of datasets are typically shown as a bounding box or polygon indicating the coverage of the underlying data (i.e., the footprint of the dataset). Multimedia information associated with particular locations (e.g. a digitized photograph of the location, text description or discussion of the location, etc.) is typically shown as icons placed on or near the location.

But there are also a number of potential problems, or requirements, for such browsing interfaces for large-scale digital library systems. One significant problem is that of clutter in the display. If all of the icons or footprints representing all of the documents in a large database associated with any geographic area visible on the digital map are shown simultaneously, the map may disappear entirely beneath a heap of icons. This largely comes down to a matter of scale in the representing the icons. If the underlying datasets are at an appropriate level of detail for the current viewing "altitude" and location of the user's display, then they should be shown, if not they should be hidden. Most GIS and some geographical browsers (for example the GIS Viewer developed for the U.C. Berkeley Digital Library project: http://elib.cs.berkeley.edu/gis/index.html) support the ability to display different datasets depending on the virtual altitude of the current display. For historical information this principle needs to be extended to the temporal dimension as well: the icons or footprints that are not appropriate for the particular temporal constraints set by the user should not be shown.

3   INDEXING HISTORICAL AND CULTURAL INFORMATION FOR GEOGRAPHIC INFORMATION RETRIEVAL

Geographic indexing and access has long been recognized as problematic in libraries (Holmes, 1990) and has not become any simpler as the libraries have become digital. In the past, indexing of geographic information in the collections of libraries has largely relied on verbal designations of places, commonly depending on the Library of Congress Subject Headings and Name Authorities as a source (Brinker, 1962; Larsgaard, 1998). Graphical methods of access (such as the use of index map sheets in cartographic collections) have been much more rare.

Text in a variety for forms and from a variety of sources comprises a significant portion of the contents of digital libraries. The sources of such text can include current information such as full-text journal or encyclopaedia articles, books, technical reports, etc. For historical collections the text may include digitization or transcriptions of manuscripts, historical published works, and commentaries on such works. For example, the Perseus project (Crane, 1996) includes (among its many collections) the full-text transcriptions of virtually all extant classical literature in the original Greek and Latin, as well as translations, commentaries on the works, lexicons and other tools. Many of the text documents in digital libraries describe, discuss, or refer to particular places or regions, events or artifacts that can be located geographically and temporally. Some of the text included in digital libraries may be, for example, names of places included in census or other data sets (see, for example, the datasets available through the History Data Service: http://hds.essex.ac.uk/) that have both geographic and temporal identification included in the data or its associated metadata.

In traditional library cataloging practice catalogers assigned geographic and temporal headings or sub-heading to items based on a judgment as to whether access by time or place would be important to users seeking that document. Digital libraries have been developed with an implicit or explicit assumption that most, if not all, of the indexing and cataloging to be performed will be done automatically (driven largely by the expense of manual indexing and cataloging). One area that has seen little work in this respect is the development of algorithms to support automatic geo-referencing of text documents. The goal is to automatically index and retrieve a document according to the geographic locations discussed, displayed, or otherwise associated with its content. Systems such as GIPSY (Woodruff and Plaunt, 1994; Larson et al., 1995) have been developed to attempt to algorithmically identify, geo-reference (and disambiguate) place names in text documents. Other work on identifying proper names in texts, including toponyms in English and other languages (Ravin and Wacholder, 1997; Kanada, 1999), has largely focused on the problems of name extraction and has not dealt with the additional issues of geo-referencing from the names themselves, or of disambiguation of location based on the texts alone. Smith and Crane (2001) have developed methods of toponym disambiguation and applied them to the needs of the large-scale Perseus digital library of historical materials. In that work Smith and Crane use a combination of statistical and heuristic methods (e.g. using nearby toponyms in a text to help disambiguate a location, as well as methods similar to GIPSY for inferring the primary location discussed in a text). Automatic extraction and disambiguation of toponyms from texts, particularly in languages other than English and for restricted temporal extents (i.e., historical periods), is still an open research area with many problems still to be resolved (see, for example Kanada, 1999).

The central issue in automatically identifying the geographic coordinates associated with places named in text documents is one of disambiguation. Even in cases where a document is meticulously manually indexed, geographic index terms consisting of text strings (such as Library of Congress Subject Headings and Name Authorities) have several well-documented problems with ambiguity, synonymy and with name changes over time (Griffiths, 1989; Holmes, 1990). Because toponyms are not unique, simple "exact match" lookup of place names is impossible. For example, San Jose and Los Angeles are common names throughout Central and South America, as well as in California and there are many London's throughout the United States and Canada. Lacking additional qualifications, it may not be possible to choose which place name is referred to in a document without further contextual information from the document or its metadata.

In historical documents, additional problems arise because the places referred to may have changed size, shape and names over time. Names of political entities are particularly fluid since political changes in the world move much faster than geological changes, and borders, country and region names, even the existence of political entities may change at any time. In addition, the names used locally for a given region may differ from common English usage, and there may be variations in the spelling of a name over time and by language (Vienna, Wien: Peking, Beijing, etc.). There are also many places described in texts that may be temporary (or temporally constrained) conventions: In historical documents, for example, names may be used by scholars to describe an area or region (study areas, battlefields, etc.) that are not part of the conventional political or geographic names of a region, but which have a definite location and temporal extent.

The basic approach to dealing with the ambiguity and transience of toponyms in digital libraries is to substitute the associated geographic coordinates (as a footprint polygon or point location) for place names in indexing and describing the documents. This does not, of course, solve the entire problem. Even though coordinates are stable regardless of name, political boundary or other changes and in spite of the fact that a location specified by coordinates is to a large extent independent of the vagaries of politics, warfare, or synonymy, for historical purposes it is often exactly those changes that are of interest to the scholar. Without the temporal dimension, closely allied to the geographic, the record of changes that are of interest to scholars might be lost.

Well-designed gazetteer resources can provide the basis for integrating digital library resources of scientific and demographic data about places with the global, multilingual records of human culture—art, literature, biography, history and other fields. Making gazetteers that support both the geographic and temporal characteristics of toponyms available to the users and contributors of digital library resources will be central to this task. Automatic georeferencing of cultural and historic information resources, primarily documents, depends on having such gazetteers available. The GIPSY system mentioned above, for example, used a gazetteer constructed from the USGS GNIS and GI-RAS databases. Today there are a number of important gazetteers being developed by a cultural, historical, and humanities projects worldwide. Because many of these are accessible over the Internet, it is becoming possible to draw on multiple, independently managed and supported, network-accessible gazetteer servers.

Unfortunately, truly effective use of gazetteers in historical and humanities computing has been impeded by the lack of standards for both gazetteer contents, and for metadata records describing the gazetteers themselves. The emerging standards for conventional gazetteer entries, based largely upon contemporary North American gazetteers that focus on contemporary mapping applications or environmental sciences are inadequate for humanities computing. Much more work is needed to extend these standards to accommodate:

- Multiple toponyms in multiple scripts and languages that refer to the same geospatial location,
- Changes and variations of toponyms over time,
- Changes over time to the boundaries, locations, and spatial footprints of places (as towns grow into cities, rivers spring their banks, and wars and revolutions redraw borders)

In addition, the range of types of geographical entities currently used in gazetteer place name type thesauri (bridge, tumulus, church) are simply not detailed enough to accommodate the range of place name types found in the global, historical texts about human culture. One of the better gazetteers and content standards in this regard is the Alexandria Digital Library (ADL) Gazetteer service and its associated content standard developed at the University of California, Santa Barbara (Hill et al., 1999). The Alexandria Digital Library Gazetteer currently contains more than 4.2 million entries with worldwide coverage. The current contents of the ADL gazetteer are based primarily on data from two US federal government gazetteers, which emphasize named features that appear on topographic maps rather than historical and cultural materials (the sources are the USGS Geographic Names Information System (GNIS) and the National Imagery and Mapping Agency's Geographic Names Processing System). The ADL Gazetteer Content Standard, however, supports many of the desirable features noted above:

"Designated name by which the entity is known and any variant names. All names can have the following optional attributes: name source, etymology, language, pronunciation, transliteration scheme, and character set. All names are flagged as current or historical and can have beginning and ending dates.

Footprints can be entered as points, lines, bounding boxes, or bounding polygons. A single gazetteer entry may have multiple footprints and each can be described as current or historical with beginning and ending dates. Measurement method, date, and accuracy can be recorded. Street addresses are also supported.

The "relationship" section of the standard supports the relationship of one gazetteer entry to another (e.g., "IsPartOf"). These relationships can be specified as current or historical and date ranges given. These explicit relationships supplement the inherent geometric and topological relationships that can be derived from the entries' footprints.

Descriptive information about the gazetteer entry is structured into (1) textual description, (2) geographic feature data, such as population, elevation, and length, (3) links (e.g., URLs) to related sources of information, and (4) a supplemental note.

Contributor Information and Source Information are described by linking to the appropriate Contributor and Source descriptive entries." (Hill et al., 1999)

The ADL gazetteer has an associated Feature Type Thesaurus, which will probably need some extensions to support all of the types of features of use in historical and cultural scholarship.

Since most existing gazetteer standards cannot accommodate the qualitative or historical information about places that constitutes a great deal of the human record of space and territory, new extensions to support references to historical records about places (which may often be incomplete, ambiguous, or conflicting) as well as the types of information included in the ADL content standard. In addition, further work is needed to develop communication protocols for communication between different digital gazetteers and to support access between gazetteers and other resources. All of the issues in searching geo-referenced digital libraries (as discussed above) apply to digital gazetteers as well.

Researchers associated with the Electronic Cultural Atlas Initiative (ECAI) are working on three necessary tasks for enabling interoperability between gazetteers, texts, and maps in a distributed environment:

1. *Standards validation, enhancement, and development*: Gazetteer content standards need to be tested on real global, historical and cultural data, and enhanced as necessary to support the international exchange of gazetteer data. Gazetteer metadata standards and protocols must be developed to allow interoperability among gazetteers and toponym-rich texts in diverse languages and formats.
2. *Infrastructure design and testing*: ECAI is working to create a multilingual union gazetteer prototype by importing XML records from several gazetteers in (initially) Chinese and English along with qualitative textual information about those places. These records will be used to establish a unicode-enabled database to link and enhance gazetteer and text records.
3. *GIS and Browser visualization*: The creation of metadata for the union gazetteer database will enable the data in the enhanced records to be viewed using the time and space visualization tools of the Electronic Cultural Atlas Initiative (ECAI) and to be linked to other globally distributed records about the same places.

This effort will be an international collaborative venture between:

1. The Electronic Cultural Atlas Initiative (ECAI) (http://www.ias.berkeley.edu/ecai). ECAI, a collaboration between IT professionals and humanities scholars, is developing a globally distributed spatio-temporal library of cultural and historical resources with a centralized metadata catalogue and a GIS viewer. The more than 250 projects under development include numerous historical gazetteer and

historical GIS projects such as the geography of Greco-Roman culture, toponym locations for over 300,000 images of Buddhist art and architecture, historical trade routes of Eurasia, the map of Hideyoshi's invasion of Korea, and historical GIS projects for China, Great Britain, the United States, the Black Sea, Tibet and other world areas.

2. Academia Sinica Computing Center (http://www.ascc.net/center/index_e.html) which is providing global access to a corpus of 2,500 years of Chinese historical writing through their Scripta Sinica project. The project currently amounts to 300 million Chinese characters. They have recently embarked on the development of a historical and contemporary gazetteer of China containing over 70,000 historical names and an additional 5,000 contemporary names) and,

3. The Alexandria Digital Library Project Gazetteer, which will provide the base gazetteer information for the project.

The proposed system, to be jointly developed between researchers at the University of California, Berkeley and the Academia Sinica, Taiwan, is intended to enable interoperability among multilingual gazetteers in diverse formats. The research will demonstrate techniques and standards that will not only enhance existing gazetteers and facilitate GIS mapping of gazetteer resources, but will also enhance capacities for linking texts, maps, and gazetteers in a distributed network architecture. Work in building entry vocabulary tools will enable the automated creation of topical indexes to place names based on multiple, diverse and multilingual authority lists.

The first steps, which we have been undertaking for the past year, concern gazetteer standards enhancement to handle multilingual issues and gazetteer metadata standards development as discussed above. This work will enable prototype linking of sample gazetteer data and texts in a multilingual environment.

Further research will be directed towards linking historical and cultural texts to the appropriate gazetteer entries, and providing disambiguated geo-references for those texts, and giving the ability to link texts with other georeferenced information. We expect that the developments in multilingual gazetteer metadata will make it possible to link the information about places found in distributed gazetteers. We hope to show that multiple gazetteers with information specific to a particular cultures, time periods, or academic disciplines (often developed by scholars for particular studies), can be linked into an integrated system. We also plan to link gazetteer data through the ECAI metadata clearinghouse to digital source materials in a variety of formats. These linkages will be bidirectional, so that a scholar may be able to trace usage of toponyms (and the resources associated with those names) in the distributed metadata, and to use GIR search methods to discover and access such text resources. In addition we hope to show that it is possible to create encyclopedic gazetteers from globally distributed records linked to qualitative information from texts.

Toponyms in these distributed gazetteers can be used to link to place names in textual materials using methods based on the GIPSY work mentioned above (Woodruff and Plaunt, 1994). This research will provide the capacity to associate toponyms with the texts in which they are mentioned. This will make it possible to link gazetteer research with ongoing developments in the creation of topical indexes and entry vocabulary searching (Buckland et al., 1999). The gazetteers that are created and linked as a result of this project will be able to be queried not only geospatially, but also topically. It will be possible for users to create second-level thematic gazetteers on the basis of texts that name places in conjunction with people, events or any other subject. The target users for the system are researchers, students, and teachers at all educational levels. The development of gazetteer reference systems linked to the qualitative information about places found in texts will enable ECAI users and project developers to conduct queries across alternative and fluid toponyms, link texts to geospatial information, and free humanities scholars with little or no training in geography from the need to determine such information for the places they study.

Geo-referenced information in the ECAI system will be viewable using ECAI's TimeMap (http://www.timemap.net) technology. TimeMap, a project of the Archae-ological Computing Laboratory at the University of Sydney is a flexible map delivery system that generates interactive time-enabled historical maps by overlaying historical datasets distributed across the Internet. TimeMap can display and animate the develop-ment of a site, a city or an empire, allowing scholars, students and teachers to discover detailed text or image information associated with any point on the map and to perform geographic analysis by overlaying information from many different digital data sources. TimeMap, which has been adopted as a core technology for ECAI, will also form an important component of the architecture proposed for the gazetteer research. A cultural and historical gazetteer system will also enhance TimeMap, providing a large global cor-pus of historical, text-linked base-map data for the other historical, cultural biographical datasets submitted to the ECAI TimeMap system and a test bed for scalability research toward future iterations of the TimeMap software.

The system envisioned for this project has a three-part architecture:

1. A gazetteer database server will house multiple sets of geo-referenced records about modern and historical places that meet a defined gazetteer record standard. Through the gazetteer database server, contributors will be able import and edit gazetteer records, and users will be able to export gazetteer records to create sub-sets or supersets. This enables and enhances development of multiple authoritative gazetteers based on reference work by individual scholars. A scholar can create a subset of gazetteer data from an authoritative gazetteer such as ADL and combine it with data from their personal research to create a specialized location infor-mation reference work. Other scholars in the same discipline can then use these specialized datasets as references for location information.
2. A gazetteer metadata clearinghouse: describing the holdings of the gazetteer database server and metadata for globally distributed gazetteers in diverse for-mats. Metadata describing the gazetteers can link these records to GIS viewers such as TimeMap. This will make it possible to visualize the gazetteer records and perform analysis by overlaying multiple gazetteers or linked datasets. This pro-vides a visualization tool for enhancing the accuracy of gazetteers. By developing a capacity to search for place names across gazetteers with conforming metadata, we will enable users to compare or map the records about places found in vari-ous gazetteer formats. This metadata will serve to link the gazetteer database with other distributed resources, by linking to a resource metadata clearinghouse such as the ECAI clearinghouse.

3. An Entry Vocabulary Index (EVI) server will draw from the gazetteer server as well as the Scripta Sinica project and other digitized texts in multiple languages. The EVI will create linkages between these different sources of information. A temporary text repository will be created for testbed purposes. By creating indexes that link gazetteer information to texts, the server will allow us to build topical indexes that are probabilistically related to particular locations as well as the inverse—location indexes which are related to particular topics. An interface will be developed to allow users to query by place names or topics and view the relevant resources.

The objective of this research is a distributed gazetteer and metadata system which links maps, texts and visual images to historical and contemporary geospatial coordinates. We believe that this system is the best solution for enabling scholars and students to join information about locations to other records of human activity. Currently, painstaking research is required each time a scholar needs to eliminate the ambiguity that arises when a single place changes name or location over time, or when a single place possesses multiple names in different languages. The gazetteer system will help to solve this problem by creating reference works of place information over time for scholars to use to link their texts, historic maps and images. This project will enrich research in gazetteer standards, computational humanities, and digital library methodology. It will also create an important reference tool that will enable scholars and students in all disciplines to better understand and use geographic information in their work.

4 PLACING EVENTS IN SPACE AND TIME: THE FUTURE OF HISTORICAL AND CULTURAL GIR

The preceding discussion has concentrated on the technology and some of the issues of in providing GIR systems for historical and cultural studies. What has not been discussed is how much of the work of scholars can be transformed by the adoption of these technologies. Some indications of what is possible can be seen by looking at some of the projects using the current generation of technologies to enhance the scholarly enterprise. As Edward Ayers, developer of the award winning "Valley of the Shadow" project (a digital library and exploration environment about two communities during the American Civil War—http://jefferson.village.virginia.edu/vshadow2/choosepart.html) has pointed out:

> "Historians at campuses all across the country, strongly encouraged by their administrations, are incorporating technology into their teaching. Students read course materials at all times of the day, talk with one another and collaborate, and embark on research projects that would have been impossible just a few years ago. Libraries, historical societies, universities, and various collaborations have created digital archives that offer new flexibility of research and exploration. The American Historical Review, Journal of American History, American Quarterly, and AHA Perspectives have devoted considerable space to professional and pedagogical changes linked to the new machinery.

As rapid as the changes have been, however, the actual writing of history has remained virtually untouched and unchanged. New technology has not affected the books and articles that form the foundation of what we teach. Other parts of the academy have sustained long-running debates over the effect of electronic media on writing, but those discussions have bypassed the historical profession almost entirely. Discussions of epistemology, narrative, and audience that have animated literary studies have had no discernable impact on historians.

The irony is that history may be better suited to digital technology than any other humanistic discipline. Changes in our field far removed from anything to do with computers have helped create a situation in history where the advantages of computers can seem appealing, and perhaps even necessary. At the same time, changes in information technology, far removed from any consideration of its possible uses for our discipline, have made it possible for us to think of new ways to approach the past. The new technologies seem tailor-made for history, a match for the growing bulk and complexity of our ever more self-conscious practice, efficient vehicles to connect with larger and more diverse audiences." (Ayers, 1999)

The Valley of the Shadow project includes such innovative uses of geo-referenced and temporal information as animated maps that show the movements of various divisions of the Union and Confederate armies as they ebb and flow from one battle to the next, crossing and recrossing the Shenandoah valley, and a complete historical GIS for Augusta County, Virginia, showing such diverse information as the concentration of slave ownership, locations of land holdings, houses, mills, etc., and fire insurance policies from 1850–1860 linked to the properties that they insure on historical maps of the area.

Other projects (such as the Perseus project, Crane, 1996) have amassed very large and complex collections of information. The Perseus Classics collection includes primary and secondary texts, site plans, digital images, and maps. It also includes art and archaeology catalogs documenting a wide range of objects: including over 1,500 vases, over 1,800 sculptures and sculptural groups, over 1,200 coins, hundreds of buildings from nearly 100 sites and over 100 gems.

And these projects are not alone, there are hundreds (if not thousands) of digital libraries (including the products of individual scholars, museums, archives and libraries) with geographically and temporally referenced information that are currently isolated from each other. Consider the impact on scholars and students alike if all of the information in these various projects could be queried, shared and combined in new ways to produce a geographical browser for all of these resources. The promise of interoperable historic and cultural digital libraries is enticing. Not only would this open new sources for scholars, but it would allow users to browse through a rich landscape of historic and cultural information and to form connections never before possible when the information sources were separated by time and space. Already, scholars are making new discoveries by the simple act of displaying their data geographically. Sometimes the discoveries of patterns in data may not be earth-shaking (such as the finding that classical Chinese scholars tended to be born along the coast, but do most of their work in the mountains or the discovery that medieval towns with shrines for local saints tended to be unaffected by the Reformation) but it is new knowledge that would not have been discovered without geographical and temporal visualization.

ACKNOWLEDGMENTS

I would like to thank Lewis Lancaster and Ruth Mostern of the Electronic Cultural Atlas Initiative for their contributions to this paper, which is derived in part from our grant applications to the National Science Foundation under the Information Technology Research and International Digital Library programs

REFERENCES

Ayers, E. (1999). The pasts and futures of digital history. http://www.vcdh.virginia.edu/PastsFutures.html.

Brimicombe, A. J. (1993). Combining positional and attribute uncertainty using fuzzy expectation in a GIS. In *GIS/LIS Proceedings*, pp. 72–81, Bethesda, MD. American Society for Photogrammetry and Remote Sensing, American Congress on Surveying and Mapping.

Brinker, B. (1962). The geographical approach to materials in the library of congress subject headings. *Library Resources and Technical Services*, 6(1):49–64.

Buckland, M., Chen, A., Gey, F. C., Kim, Y., Lam, B., Larson, R. R., Norgard, B., and Purat, J. (1999). Mapping entry vocabulary to unfamiliar metadata vocabularies. *D-Lib Magazine*, 5(1). http://www.dlib.org/dlib/january99/buckland/01buckland.html.

Crane, G. (1996). Building a digital library: The perseus project as a case study in the humanities. In *DL '96: Proceedings of the First ACM Conference on Digital Libraries*, pp. 3–10.

De Floriani, L., Marzano, P., and Puppo, E. (1993). Spatial queries and data models. In Frank, A. U. and Campari, I. (Eds), *Spatial Information Theory: A Theoretical Basis for GIS*, volume 716 of *Lecture Notes in Computer Science*. Berlin: Springer-Verlag.

ECAI (2001). Electronic Cultural Atlas Initiative. http://www.ecai.org/.

Frank, A. U. (1991). Properties of geographic data: Requirements for spatial access methods. In Günther, O. and Schek, H. J. (Eds), *Advances in Spatial Databases*, volume 525 of *Lecture Notes in Computer Science*, pp. 225–234. Berlin: Springer-Verlag.

Frew, J., Carver, L., Fischer, C., Goodchild, M., Larsgaard, M., Smith, T., and Zheng, Q. (1995). The Alexandria rapid prototype: Building a digital library for spatial information. In *ESRI User Conference Proceedings*, Redlands, CA. Environmental Systems Research Institute. http://www.esri.com/resources/userconf/proc95/to300/p255.html.

GBHGIS (2001). Great Britain Historical GIS Project. http://www.geog.port.ac.uk/gbhgis/.

Griffiths, A. (1989). SAGIS: A proposal for a Sardinian geographic information system and an assessment of alternative implementation strategies. *Journal of Information Science*, 15:261–267.

Hazarika, S. M. and Cohn, A. G. (2001). Qualitative spatio-temporal continuity. In Montello, D. R. (Ed), *Spatial Information Theory: Foundations of Geographic Information*

Science; Proceedings of COSIT 2001, volume 2205 of *Lecture Notes in Computer Science*, pp. 92–107. Berlin: Springer-Verlag.

Hill, L. L., Frew, J., and Zheng, Q. (1999). Geographic names: Implementation of a gazetteer in a georeferenced digital library. *D-Lib Magazine*, 5(1). http://www.dlib. org/dlib/january99/hill/01hill.html.

Holmes, D. O. (1990). Computers and geographic information access. *Meridian*, 4:37–49.

Jones, C. B., Alani, H., and Tudhope, D. (2002). Geographical terminology servers—closing the semantic divide. In Duckham, M., Goodchild, M. F., and Worboys, M. F. (Eds), *Foundations in Geographic Information Science*, pp. 205–222. London: Taylor & Francis.

Kanada, Y. (1999). A method of geographical name extraction from Japanese text for thematic geographical search. In *Conference on Information and Knowledge Management (CIKM '99) Proceedings*, pp. 46–54.

Larsgaard, M. L. (1998). *Map librarianship: An introduction*. Englewood, Colorado: Libraries Unlimited, 3rd edition.

Larson, R. R. (1996). Geographic information retrieval and spatial browsing. In Smith, L. and Gluck, M. (Eds), *Geographic Information Systems and Libraries: Patrons, Maps, and Spatial Information*, pp. 81–124. Urbana-Champaign: University of Illinois, GSLIS.

Larson, R. R., Plaunt, C., Woodruff, A. G., and Hearst, M. A. (1995). The Sequoia 2000 electronic repository. *Digital Technical Journal*, 7(3):50–65.

Laurini, R. and Thompson, D. (1992). *Fundamentals of Spatial Information Systems*. San Diego, CA: Academic Press.

Peuquet, D. J. (2001). Making space for time: Issues in space-time data representation. *GeoInformatica*, 5(1):11–32.

Ravin, Y. and Wacholder, N. (1997). Extracting names from natural-language text. Technical report, IBM Research Division. IBM Research Report RC 20338 (04/10/97).

Salton, G. (1989). *Automatic Text Processing: The Transformation, Analysis, and Retrieval of Information by Computer*. Reading, MA: Addison-Wesley.

Smith, D. A. and Crane, G. (2001). Disambiguating geographic names in a historical digital library. In Constantopoulis, P. and Sølvberg, I. T. (Eds), *ECDL 2001 Proceedings*, pp. 127–136, Berlin. Springer-Verlag.

Woodruff, A. G. and Plaunt, C. (1994). GIPSY: Automated geographic indexing of text documents. *Journal of the American Society for Information Science*, 4(9):645–655.

CHAPTER 13

Geographic Activity Models

Sabine Timpf
Department of Geography, University of Zurich
Zurich, CH-8057, Switzerland

1 MOTIVATION

The main impediment to a widespread use of geographic information using a geographic information system (GIS) is the gap between the user's expression of a task within the context of an activity and the sequence of operations within a GIS needed to successfully perform that task. Users express their intentions and supporting activities in a high level language, usually in natural language. This level corresponds to the knowledge level (Newell, 1982). Those activities need to be broken down into a sequence of tasks that exist in the head of the user and are part of her training as expert in the field. Each task has to be performed as a sequence of operations (according to a specific method), which can be carried out (mostly) by the commands and functions of an information system, e.g., a GIS or a Spatial Decision Support System. Depending on the complexity of the activity, more intermediate levels of tasks might be needed, creating sub-tasks.

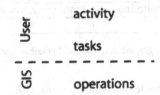

Figure 1: Division of work between user and GIS

The high-level language is part of the user or application domain, whereas the low-level operations are part of the system domain (Figure 1). At the moment the user is in charge of translating between the three levels and creating the intermediate levels. In our opinion at least part of this translation can be done automatically with the help of geographic activity models. Instead of having to deal with operations, the user should deal with the methods needed to carry out the task within the framework of the activity (Figure 2).

241

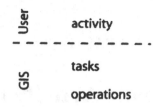

Figure 2: Desired division of work between user and GIS

We propose to research geographic activities and corresponding methods in geographic information processing (Table 1). We will discuss questions such as: What are the tasks that we need to solve a specific problem within an activity? What are the methods that we use to accomplish a task? Which are the operations that we need within a GIS for a specific method? Can we describe those methods (not the operations within a GIS) such that they can be re-used by others? We argue that the analytical potential of GIS can better be exploited once the user is provided with significant choices to adapt the computing environment to the activity at hand.

Geographic Activities	Methods
Site location	spatial analysis, visualization
Wayfinding	planning, simulation
Utility management	inventory, decision-support
Emergency management	decision-support, inventory
Landscape conservation	inventory, monitoring
Urban planning	decision-support, planning
Environmental monitoring	monitoring, visualization
...	

Table 1: Some geographic activities and possible associated methods

We propose to describe problem-solving methods, task ontologies and sequencing in geographic activity models (GAM). This work is a discussion of such models, what they are, how they are supposed to work, and how they should be derived. Geographic activity models should be conceived of as plug-in modules that adapt an information system to a specific task. They represent a fundamental change in the way to operate an information system. They offer problem-solving knowledge to the information system and thus can transform the vocabulary of the generic information system to that of an application-specific information system. The idea is to make current geographic information systems more intelligent, i.e., to include knowledge about common problem-solving processes, and to enable the reuse of problem-solving methods.

Section 2 reviews previous and related work. Section 3 introduces geographic activity models in detail. Section 4 discusses geographic activity models in the context of three different research perspectives (usability, interoperability, semantics). Finally, section 5 concludes and presents future work.

2 PREVIOUS AND RELATED WORK

Examination of the functionality of GIS (Maguire and Dangermond, 1991; Albrecht, 1997) has shown that current GIS are designed to handle data and enable reuse of data. Maguire and Dangermond (1991) analyzed high level generic tasks in GIS. They identified data capture, data transfer, data validation and editing, storing and structuring data, restructuring data, generalizing data, transforming data, querying data, analyzing data, and presenting data as generic tasks of a GIS. Those tasks are generic data handling tasks and this categorization provides a first grouping of operations into tasks. However, the phrase "analyzing data" hides the application-specific knowledge necessary to extract meaning from data.

Albrecht (1996, 1997) presents 20 universal GIS operations derived from questionnaires out of a compiled list of 144 GIS analytical operations and functions from diverse GIS. The operations are listed in Table 2.

Search	Interpolation	Spatial Search	Thematic Search	Reclassification
Locational Analysis	Buffer	Corridor	Overlay	Thiessen/ Voronoi
Terrain Analysis	Slope/ Aspect	Watershed	Drainage/ Network	Viewshed
Distribution/ Neighborhood	Cost/ Diffusion/ Spread	Proximity	Nearest Neighbor	
Spatial Analysis	Multivariate Statistics	Patterns/ Dispersion	Centrality/ Connectedness	Shape
Measurements	Measurement			

Table 2: Universal GIS operations (Albrecht, 1997)

Those operations are derived from the user perspective although heavily influenced by the available operations in GIS. Those operations are not the tasks we are looking for. Research will show if those 20 operations indeed represent the building blocks of all geographic tasks within a certain application area. As Albrecht (1997) remarked, a good GIS user interface needs to adjust to the field of application, i.e., the categorization of the operations depends on the application. This is expressed, for example, in the group heading and the names of operations within the groups, which should conform to the commonly accepted knowledge within the user group of a specific application. There need to be more research carried out to derive this kind of knowledge. In part this knowledge can be taken from accounts of customization procedures of GIS.

The incorporation of domain-specific problem-solving knowledge has been promoted by research on Intelligent GIS (Birkin et al., 1996). The authors discuss a combination of conventional GIS and model-based analysis. The incorporation of reasoning mechanisms and knowledge bases into current GIS to make GIS more intelligent is also subject to research by Yuan (2001). The aim in this project is to support spatiotemporal queries, analysis, and modeling in hydrology.

Kuhn criticises that GIS do not support human activities Kuhn (2001). His "domain theories are based on the assumption that the increasing complexity of human conceptualizations of the environment results from [...] increasingly complex activities, rather than the other way round" (p.28). Along the same lines Câmara et al. (2000) promotes the idea of action driven ontologies. Both authors appear to adopt the paradigm that activities (or actions) drive the conceptualization of the domain of study. We infer from their work that more knowledge about geographic activities is necessary in order to study the influence of activities (actions) on human conceptualizations of the geographic world and problem-solving behavior.

At the NCGIA specialist meeting on user interfaces the need for research on typology of users, use types and GIS tasks to improve usability of GIS was identified (Mark and Frank, 1992; Mark, 1994). Geographic activity models provide a framework in which to encode knowledge about GIS tasks, their goals, constraints, and their users.

Within the knowledge engineering community Chandrasekaran (1986) has brought up the concept of universality with the idea of generic tasks. His aim was to model the problem-solving process in medical diagnosis. He showed that diagnosis consists of several generic methods and defined those methods formally. Subsequent work dealt with the application of those generic problem-solving methods to other domains. We believe that for certain geographic activities similar generic methods exist. Geographic activity models are a means to store those methods such that they can be used by expert and novice GIS users alike.

Bucher (2000, 2001, 2002) is working on a geographic task server to provide users with additional knowledge on how to manipulate data. The information is structured according to a task-method-tool model. This work relies on the CommonKADS framework (Schreiber et al., 1994) for describing problem-solving knowledge. The goal of Bucher's work is an expert system that stores patterns of manipulation of geographic data. There is but a small step from patterns to problem-solving knowledge and from there to geographic activity models.

3   GEOGRAPHIC ACTIVITY MODELS

A geographic activity model contains a (formal) description of a task and states its goal or goals. It names the data requirements (input) and the results (output) of the task. It also contains a description of its parameters in the space of task, i.e. an indication of its standing in relation to other tasks. It lists possible methods to carry out the task. It gives an account of the necessary subtasks, i.e., the task chain. It describes how the information flows between the subtasks in the task chain. Finally, it defines possible constraints (e.g., quality) on data, methods, input, and results.

Geographic activity models can be conceived of as plug-in modules that adapt an information system to a specific task. For example, the user interface is adapted by showing under a specific command the necessary steps (or subtasks) to carry out the task. Depending on the skill of the user more or less detail on the subtasks (information on operations, methods, constraints, and data requirements) is given. Even the sequence of those steps could be subject to change, i.e., an expert user can modify the list of subtasks (Brazier et al., 2000), whereas a novice user will operate with a given sequence. This classification corresponds to the distinctions between types of task-chaining: opaque task chaining does not disclose detail to the user, translucent task chaining requires user interaction at certain points, whereas transparent task chaining leaves the chaining of tasks up to the user.

Geographic activity models are more than an adaptation of the user interface. They represent a fundamental change in the way to operate an information system. They plug problem-solving knowledge into the information system and thus transform the vocabulary of the generic information system to that of an application-specific information system. This changes the possible interpretations of the data, which might contribute to the solution of the semantics problem. At the same time necessary methods for the solution of the task are identified which can be accessed over the Internet. A feasibility check of the data is executed to derive possible problems with input or output of the subtasks and to check if the constraints of the task are fulfilled.

There are four ways to derive geographic activity models. The first one is a task-analysis, i.e., an analysis of the subtasks, operations, and data used in solving a specific problem (see for example Raubal, 1997; Timpf et al., 1992; Timpf, 2001). Secondly, information on past task-analyses should be available in the literature on GIS applications. During this literature search it would be beneficial to classify tasks per application. This would help in extracting the methods used to solve certain tasks and to identify domain-independent strategies. Thirdly, customized products should show an alteration of the user interface and an adaptation of the methods and operations used, which are observable and attributable to specific tasks. Finally, we can apply results from knowledge engineering research to geographic tasks.

3.1 Geographic problem-solving methods

A Problem-Solving Method (after McDermott, 1988) is an abstract model of problem-solving with the following components:

- Actions that accomplish tasks, expressed in a behavioral way
- Recursive decomposition into subtasks, solved by another method until mechanisms.
- Selected w.r.t. factors (e.g. availability of data; time, space and quality requirements)

Typically problem-solving methods are specified in a task-specific fashion, using modeling frameworks which describe their control and inference structures as well as their knowledge requirements and competence (Fensel et al., 1997). Describing problem-solving methods in the style of CommonKADS (Schreiber et al., 1994) requires specification of much of the internal reasoning process of a problem-solving method. In particular, the following descriptions need to be given:

1. the internal reasoning steps of the problem-solving method;
2. the data flows between the reasoning steps;
3. the control that guides the dynamic execution of the reasoning steps;
4. the knowledge requirements of a problem-solving method;

5. the goals that can be achieved by a problem-solving method.

However, most of these aspects have to do with understanding how a problem-solving method achieves its goals. To assess the applicability of a problem-solving method one only needs knowledge about its competence and domain requirements—i.e. (4) and (5) above.

The difference between a geographic activity model (GAM) and a problem-solving model is that GAMs provide the user with a choice of methods to solve her problem. The choice might depend on the type and quality of available geographic data and a GAM will present criteria for the user to decide which method to use with which data.

3.2 The method hyperspace

The method-hyper-space (MHS) is a formal description of the problem solving methods used in geographic information processing. It is in fact an n-dimensional space where the axes denote the parameters that determine which method to use when. At the moment it is a hypothesis that such a set of independent parameters, determining the use of a method, exists.

Each method occupies a region within this hyperspace. Those regions can overlap, meaning that both methods can be used for at least one specific problem type. If those regions are disjoint, then the two methods cannot solve the same problem type. Perhaps more topological relations are meaningful.

Given a specific problem/task the method-hyperspace shows which methods or combination of methods can possibly be used to solve this problem. The final decision is also dependent on the available data and on performance and optimization criteria.

Some methods might be independent of the application domain, others are clearly dependent on the domain. Can we make a difference, are independent methods more generic than dependent methods? Can both types of methods be described in the same framework, e.g., the one as a specialization or instantiation of the other (as in, e.g., Brazier et al., 1995)?

3.3 A simple informal example

This example is taken from an introductory course on GIS. The task is to derive ideal sites for a villa, where the ideal site is determined by several criteria. The task chain including subtasks and operations is given in Tables 3 and 4. In general the task chain is as follows (see also Figure 3):

- determine criteria 1..n (user input)
- determine locations for criterion 1–n (spatial analysis)
- determine joint locations (spatial analysis)
- map locations (visualization)

The spatial data used are raster data with the information for each criterion. The data flow in this example is rather simple. For each criterion possible locations are determined independently and later combined (for specific operation chains please refer to Table 3). The base data is a digital terrain model, information on ground cover (vector transformed to raster), and three locations that should be seen. As special input the criteria are given. The output of the task is a set of possible locations (in grid cells).

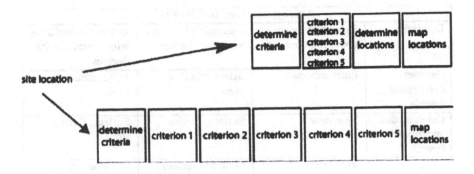

Figure 3: Task chain for site location

As shown in Figure 3 there are two possible ways to determine a site from a list of criteria. In the upper task chain we assume that the criteria are independent from each other. Thus they can be computed in parallel. In the lower task chain the criteria are applied sequentially to the result set of the previous computation. This results in less data to hand over to the next task step and reduces the time for computation drastically. Both methods have drawbacks and advantages. It may depend on system resources which solution is chosen.

4 DISCUSSION

Geographic activity models have three main objectives: make GIS more usable and adaptable (i.e., the usability issue), enable geographic information brokering (i.e., the interoperability issue), and provide the context for data (i.e., semantics issue).

4.1 Improving the usability of GIS

Users of current Geographic Information Systems (GIS) are experts in their own domain. They are interested in solving their problem, planning and designing, simulating future scenarios, assessing a risk, mapping, or getting help in making a decision. A geographic information system is to the users, for example,

- a visualization tool,
- an inventory tool, (acquire and present)
- a decision-support tool, (all types)
- a spatial analysis tool, (cover and differentiate)
- a simulation tool, (propose and revise, extrapolate from a similar case
- an intelligent planning tool, (propose and revise, extrapolate)

Activity: Site Location (Part 1)			
Determine criteria	Determine locations above fog boundary	Determine locations with evening sun	Determine locations with slopes <= 25 degrees
Get user determined criteria	Enter arcview	Select dtm25 in view	Select dtm25 in view
Get data: dtm and ground cover	New view	Surface, Derive aspect	Surface, Derive slope
	View, view properties: give the view a name	Look for number interval southwest-west: 202.5–292.5	MC slope_dtm25 <= 25
	Load spatial analyst	MC aspect_dtm25 >= 202.5 and aspect_dtm25 <= 292.5	
	Add theme: grid dtm25		
	Select in view dtm25		
	Map calculator (MC) dtm25 < 550		
	Legend of calc1: change foreground and background color of class 0=False, do not show NoData;	Legend of calc1: change foreground and background color of class 0=False, do not show NoData;	Legend of calc1: change foreground and background color of class 0=False, do not show NoData;
	Theme, theme properties: change name in legend; theme, save dataset as: give it a proper name	Theme, theme properties: change name in legend; theme, save dataset as: give it a proper name	Theme, theme properties: change name in legend; theme, save dataset as: give it a proper name

Table 3: Task and operation chains of the activity site location

Activity: Site Location (Part 2)			
Determine locations with a view on the three given locations	Determine locations not in forest and not built-up	Determine locations that satisfy all those criteria	Map possible locations
View, new theme, points	add theme, ground_cover	MC, villaLocations = notfog and eveningSun and flatSlope and goodView and groundcover	add theme, topographic map
select given locations	open table, identify classes		select villaLocations, legend: color red
stop editing, save theme	Query Builder, dxf=other (not forest, not built-up, not vineyards, not lake)		print or show on screens
open table, deselect points	Theme, convert to grid		
select dtm25 and locations in view			
Surface, calculate Viewshed			
legend of calc1: change foreground and background color of class 0=False, do not show NoData;	legend of calc1: change foreground and background color of class 0=False, do not show NoData;	legend of calc1: change foreground and background color of class 0=False, do not show NoData;	
theme, theme properties: change name in legend; theme, save dataset as: give it a proper name	theme, theme properties: change name in legend; theme, save dataset as: give it a proper name	theme, theme properties: change name in legend; theme, save dataset as: give it a proper name	

Table 4: Task and operation chains of the activity site locations

- a design tool
- or a combination of the above.

(For a similar list see also Breuker, 1994.)

The current generation of GIS does not live up to this image. Users must have knowledge about the intricacies of dealing with GIS operations in addition to their own field of expertise. They cannot concentrate on 'doing their job' wielding a powerful tool. This greatly reduces the usability of GIS and also the value of GIS, because value is only derived from geodata by use (Krek and Frank, 2000).

The list given above presents user activities in increasing complexity although no total order is implied. But it suggests that a spatial problem solver might be more complex than a spatial decision-support tool, which in turn is more complex than a mapping tool. This has implications for the organization of a GIS. If the activities can be described such that each step or combination of steps can be represented by a module in a GIS, then the GIS can be tailored to the activity by providing exactly those modules that allow the user to carry out their activity. A GIS should also be adaptable: depending on the knowledge and skill of the user it will present more or less functions, apart from tailoring the shown functions to the application at hand (Davies and Medyckyj-Scott, 1994).

The current GIS functionality has two distinct markets with similar consequences for the 'inner working' of the GIS. Within the mass market (e.g., location-based services), a user will be completely unaware that she just processed spatial information. Within the expert market, the GIS will blend into the background and put the focus on the task at hand. The expert might be more interested in the methods and algorithms that are used for her specific application area. To satisfy this user a greater transparency in the use of tasks and methods is needed. However, in both cases the non-task-essential computing processes become invisible (Norman, 1998), leaving the focus to the task at hand. We believe that this invisibility of non-task-related computing processes can be achieved by using geographic activity models.

4.2 Interoperability: Sharing information—sharing methods

Interoperability deals with sharing information that is distributed over different platforms, geographic locations, and database systems (Goodchild et al., 1999). But it also deals with sharing and accessing distributed services, i.e., methods or programs. One of the main challenges for interoperable GIS is the sharing of semantic information and the intelligent reuse of services. GIS now deal with the management and storage of data for reuse, in the future they will also have to deal with the reuse of methods and tasks descriptions.

In recent years two main technologies for knowledge sharing and reuse have emerged: *ontologies* (Gruber, 1993) and *problem solving methods*. Ontologies specify reusable conceptualizations, which can be shared by multiple reasoning components communicating during a problem solving process. Problem-solving methods describe in a domain-independent way the generic reasoning steps and knowledge types needed to perform a task (Fensel et al., 1997).

Scenarios in interoperability (Kottman, 2000) rely on the existence of software that can deal with redirecting queries to appropriate addresses for handling or computing and sending the compiled answer back to the inquirer. This software is called the information broker. The information broker (Timpf, 2001) is in charge of *task chaining*, i.e., breaking the query down into a sequence of tasks and sending tasks or subtasks off for computation (see also Fensel, 1997). Task chaining requires knowledge about possible task decompositions of the query (Yang, 1997; Brazier et al., 1995), i.e., exactly that knowledge which we intend to provide with a geographic activity model. Geographic activity models contain knowledge about the task sequence, the task hierarchy, and constraints to the computation process. Thus an information broker would directly benefit from the knowledge embedded in a geographic activity model.

4.3 Tasks provide context for data

A task guides cognition and perhaps even perception of objects in a given situation. The reason for a task and the way we perform this task guide which parts of reality we look at and perceive. For example, we give different route directions to a pedestrian than to a truck driver: our cognitive model of the route changes with the specific task. The task influences the types of objects and the parts of objects that we consider important, i.e., object ontology and its level of abstraction. The directions for a pedestrian yield different objects (sidewalks, foot-paths, stairs, etc.) than the directions for the truck driver (highways, stop lights, one-way streets etc.).

Tasks produce partitions of reality (Smith, 2000), where reality is composed of those things that are interesting (the smaller part but very detailed) and of those things that are not interesting for the task at hand (the larger part). *Domain ontologies* (in the sense of Guarino, 1998) describe those parts of partitions, i.e. concepts, which are interesting for a certain domain. These concepts are used for all tasks that occur within that domain. *Problem-solving methods* describe the reasoning concepts and their relationships occurring for a specific task (see above).

Any time we use data for a specific purpose, data is metamorphosed into information. The emphasis in GI science research so far has been on data—how to represent, how to measure, visualize geographic data. We do not know much about the tasks this data is used for although tasks seem to play a big role in determining the meaning of data. From this observation we infer a need to do research about geographic tasks.

The current hypothesis is that tasks provide context for data and thus solve the semantic ambiguity problem of data. If we were able to describe data sets in combination with associated tasks and knew formal relations between tasks, we would be able to tell if the data used for task 1 can also be used for task 2. If task 2 is more specific, we will need additional or more detailed data to solve it. If the task 2 is more generic, then we need abstraction mechanisms to abstract from the existing data. If the tasks are at the same level of abstraction, then the question is if they share a common generic task. If they do, the likelihood increases that the knowledge used for the first task can be re-used for the second.

5 CONCLUSIONS: A NEW PARADIGM?

The main impediment to a widespread use of geographic information using a geographic information system (GIS) is the gap between the user's expression of a task within the context of an activity and the sequence of operations within a GIS needed to successfully perform that task. This corresponds to the gap between what Newell (1982) calls the knowledge level and the symbolic (computational) level. It is our conviction that GIS need to communicate with the user at the knowledge level. One possibility to do so is to include information about problem-solving methods, user activities and task chains within a geographic information system, thus making the system more intelligent and usable. The idea is to provide the user as often as possible with knowledge level information in the form of generic tasks and methods instead of requiring her to learn specific GIS operations.

Tasks are the units of work in which people think, activities provide the motive behind a task chain, and sequences of operations carry out the tasks without being themselves goal-directed. Within a distributed, interoperable environment tasks need to be coordinated and then chained for completion. The idea is to reuse knowledge and problem-solving methods as often as possible. Knowledge about task chaining is inherent in problem-solving methods. Unfortunately, we do not have a formalized body of knowledge about geographic problem-solving methods and task chaining.

This research presents a model called geographic activity model to store information about user activities, task chains and corresponding operations, and problem-solving knowledge. Geographic activity models will:

- Adapt user interface to a specific task
- Choose and plug in the problem-solving knowledge i.e. the task chain
- Enable interoperability: knows about the location and quality of necessary methods (subtasks included)
- Disambiguate the semantics of the domain knowledge (i.e. data).

We have shown using a simple site location example how problem-solving knowledge can be extracted and how different methods arrive at the same result. Bucher (2002) has created a framework in which to store this knowledge. More analysis of problem-solving methods and associated activities, tasks, and operations is necessary to complete this knowledge base. This endeavour cannot be the task of a single research group. Therefore, we are proposing to deal with problem-solving knowledge with the same intensity as we have been dealing with data issues in the past. This requires a shift of paradigm from emphasizing what-knowledge to emphasizing how-and why-knowledge.

REFERENCES

Albrecht, J. (1996). Universal GIS operations—a task-oriented systematization of data-structure independent GIS functionality leading towards a geographic modeling language. In *ISPA*, Vechta. University of Vechta.

Albrecht, J. (1997). Universal analytical GIS operations. In Craglia, M. and Onsrud, H. (Eds), *Geographic Information Research: Trans-Atlantic perspectives*, pp. 577–591. London: Taylor & Francis.

Birkin, M., Clarke, G., Clarke, M., and Wilson, A. (1996). *Intelligent GIS: Location decisions and strategic planning*. Cambridge: Pearson Professional Ltd.

Brazier, F. M. T., Jonker, C. M., Treur, J., and Wijngaards, N. J. E. (2000). On the use of shared task models in knowledge acquisition, strategic user interaction and clarification agents. *International Journal of Human-Computer Studies*, 52:77–110.

Brazier, F. M. T., Treur, J., Wijngaards, N. J. E., and Willems, M. (1995). Formal specification of hierarchically (de)composed tasks. In *9th Banff Knowledge Acquisition for Knowledge-Based Systems Workshop KAW-95*, Calgary. SRDG Publications.

Breuker, J. (1994). Components of problem solving. In Steels, L., Schreiber, A. T., and de Velde., W. V. (Eds), *A Future of Knowledge Acquisition*, pp. 118–136. Heidelberg: Springer-Verlag.

Bucher, B. (2000). Users access to geographic information resources. In Winter, S. (Ed), *Geographical Domain and Geographical Information Systems*, pp. 29–32. Vienna: Institute for Geoinformation, Vienna University of Technology.

Bucher, B. (2001). A model to store and reuse geographic application patterns. In Konecny, M. (Ed), *AGILE—GI in Europe: Integrative Interoperable Interactive*, pp. 289–295. Brno, Czech Republic: Masaryk University Brno.

Bucher, B. (2002). A geographic tasks server—implementation of the locating task. GIS-RUK 2002, Sheffield, University of Sheffield.

Câmara, G., Montiero, A. M. V., Paiva, J. A., and de Souza, R. C. M. (2000). Action-driven ontologies of the geographical space: Beyond the field-object debate. In *GI-Science 2000*, Savannah, GA.

Chandrasekaran, B. (1986). Generic tasks in knowledge-based reasoning: High-level building blocks for expert system design. *IEEE Expert*, 1(3):23–30.

Davies, C. and Medyckyj-Scott, D. (1994). GIS usability: Recommendations based on the user's view. *International Journal of Geographical Information Systems*, 8(2):175–190.

Fensel, D. (1997). An ontology-based broker: Making problem-solving method reuse work. Workshop on Problem-Solving Methods during the IJCAI, Japan.

Fensel, D., Motta, E., Decker, S., and Zdrahal, Z. (1997). Using ontologies for defining tasks, problem-solving methods and their mappings. In Plaza, E. and Benjamins, V. R. (Eds), *Knowledge Acquisition, Modeling and Management*, pp. 113–128, Berlin. Springer-Verlag.

Goodchild, M. F., Egenhofer, M. J., Fegeas, R., and Kottman, C. A. (Eds) (1999). *Interoperating Geographic Information Systems*. Dordrecht: Kluwer Academic Publishers.

Gruber, T. R. (1993). What is an ontology? http://www-ksl.stanford.edu/kst/what-is-an-ontology.html.

Guarino, N. (1998). Formal ontology and information systems. In *Formal Ontology in Information Systems (FOIS)*, Amsterdam. IOS Press.

Kottman, C. (2000). The OpenGIS abstract specification. Topic 17: Location-based/mobile services. Technical report, Open GIS Consortium.

Krek, A. and Frank, A. U. (2000). The economic value of geo information. *Geo-Informations-Systeme—Journal for Spatial Information and Decision Making*, 13(3):10–12.

Kuhn, W. (2001). Ontologies in support of activities in geographical space. *International Journal of Geographic Information Science*, 15(7):613–631.

Maguire, D. J. and Dangermond, J. (1991). The functionality of GIS. In Maguire, D. J., Goodchild, M. F., and Rhind, D. W. (Eds), *Geographical Information Systems: Principles and applications*, volume 1, pp. 319–335. Essex: Longman Scientific & Technical.

Mark, D. M. (1994). Research initiative 13: User interfaces for geographic information systems. Technical report, NCGIA, Buffalo, USA.

Mark, D. M. and Frank, A. U. (1992). NCGIA research initiative 13. report on the specialist meeting: User interfaces for geographic information systems. Technical Report 92-3, NCGIA, Santa Barbara.

McDermott, J. (1988). Preliminary steps toward a taxonomy of problem-solving methods. In Marcus, S. (Ed), *Automating Knowledge Acquisition for Expert Systems*, pp. 225–256. Boston: Kluwer.

Newell, A. (1982). The knowledge level. *Artificial Intelligence*, 18:87–127.

Norman, D. A. (1998). *The Invisible Computer*. Cambridge, MA: MIT Press.

Raubal, M. (1997). *Structuring Wayfinding Tasks with Image Schemata*. PhD thesis, Department of Spatial Information Science and Engineering, University of Maine, Orono, ME.

Schreiber, A. T., Wielinga, B., and Van de Velde, W. (1994). CommonKADS. a comprehensive methodology for KBS development. *IEEE Expert*, 9(6):28–37.

Smith, B. (2000). Ontological imperialism. Talk held at GIScience 2000, Savannah, GA.

Timpf, S. (2001). The information broker: Problem-solving knowledge for location-based services. In Konecny, M. (Ed), *AGILE—GI in Europe: Integrative Interoperable Interactive*, pp. 203–204. Brno, Czech Republic: Masaryk University Brno.

Timpf, S., Volta, G. S., Pollock, D. W., and Egenhofer, M. J. (1992). A conceptual model of wayfinding using multiple levels of abstractions. In Frank, A. U., Campari, I., and Formentini, U. (Eds), *Theories and Methods of Spatio-Temporal Reasoning in Geographic Space*, volume 639 of *Lecture Notes in Computer Science*, pp. 348–367. Berlin: Springer-Verlag.

Yang, Q. (1997). *Intelligent Planning—a decomposition and abstraction based approach*. Berlin: Springer-Verlag.

Yuan, M. (2001). Development of an intelligent geographic information system to support spatiotemporal queries, analysis, and modeling in hydrology. http://ouucgis.ou.edu/nima_abs.html.

Index

Printed in the United States
by Baker & Taylor Publisher Services